Liz Hope 89

PENGUIN
SELF-
STARTERS

Basic Statistics

Peter Gwilliam was educated at Sussex University and is
currently Head of Mathematics at a comprehensive school in
Lewes, Sussex.

ADVISORY EDITORS: Stephen Coote and Bryan Loughrey

Basic Statistics

Peter Gwilliam

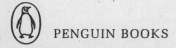

PENGUIN BOOKS

PENGUIN BOOKS

Published by the Penguin Group
27 Wrights Lane, London W8 5TZ, England
Viking Penguin Inc., 40 West 23rd Street, New York, New York 10010, USA
Penguin Books Australia Ltd, Ringwood, Victoria, Australia
Penguin Books Canada Ltd, 2801 John Street, Markham, Ontario, Canada L3R 1B4
Penguin Books (NZ) Ltd, 182–190 Wairau Road, Auckland 10, New Zealand

Penguin Books Ltd, Registered Offices: Harmondsworth, Middlesex, England

First published 1988

Typeset, printed and bound in Great Britain by
Hazell Watson and Viney Limited
Member of BPCC plc
Aylesbury Bucks
Typeset in Linotron 202 Melior

For my father,
Dr A. E. Gwilliam,
a great mathematician and teacher.
He was an inspiration to many.

Contents

8   Contents

go over

go over

Contents

Preface

Statistics, as a subject, is becoming increasingly important in our lives. We are presented every day with the results of polls and surveys which are intended to influence as well as reflect our opinions. In business life the proliferation of computer software which does wonderful things with numbers makes it essential to have a more than passing acquaintance with the theory behind the calculations. The social sciences – Psychology, Sociology, Economics, etc. – rely increasingly on the use of Statistics.

This book is intended for those with little or no understanding of Statistics and with reservations about their mathematical ability. Starting with very few assumptions about mathematical background (just a familiarity with the four arithmetic buttons on a calculator), it provides the basic ideas and much of the vocabulary of Statistics.

Most Statistics texts cover this groundwork in the first few chapters and then gallop on. The idea of *Basic Statistics* is to provide a more gentle approach to the topic. Whether you are studying the subject for GCSE, or as a component of some other course or are just looking for a general grounding, this text should suit you.

Such a book cannot hope to penetrate the subject to any great depth. I have chosen to concentrate, therefore, mainly on descriptive statistics. This is the starting place for any further work on using statistics to draw inferences and make predictions. I have in some places pointed to the directions possible, but by and large I have left such advanced work to other texts, of which there are many. *Basic Statistics* will give you everyday competence and open the way for further reading and study.

Nowadays no one needs to do other than simple calculations without a calculator. I assume throughout that you will be using one. Even then, much of the number crunching associated with the subject is of a highly repetitive nature, lending itself readily to the use of a computer. An appendix at the back offers some programs in BASIC which will do some of this work for you.

The book contains a number of simple and, I hope, not too daunting exercises. The best source of exercises, however, is the environment in which you find yourself working or studying. Statistics is an applied subject and should at all times be related to real situations.

Having said that, all characters contained within this book are entirely fictitious and bear no relation or resemblance to any person, country or place.

1 Introduction

'Knowledge is power.' Francis Bacon's adage is especially relevant to the modern world. With the aid of computers we are producing and publishing more and more information every day and hardly any aspect of our lives is left undocumented. The danger is that we are about to be swamped by facts and figures, which only 'experts' will be able to understand and control.

Statistics fills the crucial gap between information and knowledge. As a discipline it provides us with tools to condense and make sense of what we observe and record about the world around us.

The idea of gathering and processing information for purposes of gaining or maintaining power is very old. A major contributory factor to the success of the Roman empire was their ability to deal with information. Remember that the birth of the Christian era coincided with the collection of statistical data.

The word 'statistics', however, was not used until much later, in Bacon's age, when the increasingly complex nature of the seventeenth-century state required the attentions of 'statists'. These were specialists in those aspects of running a state which were particularly related to numbers. This encompassed the tax liabilities of the citizens as well as the state's potential for raising armies.

The first usage of the actual name 'Statistics' was in 1672 although it did not acquire its modern sense until the early eighteenth century. The father of modern statistics was Sir William Petty, a seventeenth-century polymath and statesman, who called it 'political arithmetic'.

'By political arithmetic we mean the art of reasoning by figures upon things relating to government' (*Sir William Petty, 1623–1687* by Lord Edmond Fitzmaurice, published in London in 1895, p. 183). He used it to try to provide rational arguments for his belief that England was not in such a bad state as people were making out and was certainly no worse off than France, where the court of Louis XIV was dazzling Europe. *Plus ça change!*

Since then the discipline has found many areas of application beyond government. It has progressed dramatically in scope, finesse and power but still handles mainly numerical data. These figures are now called statistics after the subject, Statistics, devoted to making sense of them.

In this book we look primarily at the processing rather than the gathering of statistics. This is the field where the mathematician holds sway and is, as a consequence, the area where most people have problems. I have tried in this text to keep the complicated mathematics in the background. The emphasis is mainly on *how* we do statistics and not on *why* we do it like that. As a consequence I will ask you to accept a fair amount on trust. If you really want all the formal proofs there are plenty of books around which provide them.

Since statistics is, by its nature, based on mathematics or, more particularly, arithmetic, it is inevitable and unavoidable that this book contains a fair amount of both. Where these threaten to become intrusive I have moved them either to the end of the chapter or the end of the book. All that is required of you, gentle reader, is a reasonable familiarity with a calculator and a passing acquaintance with the language of graphs.

It may seem perverse to place the chapter on collecting data after all those based on processing it. I have chosen this arrangement deliberately since it is all too easy to design a survey or an experiment without thinking about the nature of the results and the way in which we are going to process them. The type of manipulation and the subsequent interpretation will significantly affect the way in which we gather our information.

I have placed in the last chapter some of the more arcane areas of arithmetic such as place value, absolute and relative error and significant figures. The book can easily be read without reference to these although ultimately a familiarity with them will improve our statistical understanding. Where numbers are quoted as the results of calculations you may well find, if you check, that they differ from those provided by your calculator. I have not gone wrong (I hope!) but rather I have shortened answers where appropriate. Because I have done this consistently, I have not, in general, bothered to note it except in the section specifically explaining how to do it.

In keeping with the title of the book I have included several computer programs written in BASIC which illustrate or make clearer some of the areas covered.

Despite Mark Twain's warning about 'Lies, Damned Lies and Statistics', the subject, and the data which it produces, are often given more credence than they deserve. This is probably due to the mathematical connection. Most people hold mathematics in such respect and yet know so little about it that it can impart spurious respectability to otherwise shady endeavours.

An acquaintance of mine, and, I suspect, of yours, goes by the name of GUS (also known as the Generally Unscrupulous Statistician). He plays on this trust in many ways. During the course of the book we will see him in action and also gain the necessary knowledge to recognize and refute his handiwork.

By the end of the text you should feel more confident with the terminology and some of the simpler methods and ideas associated with this area of information processing.

I have chosen to deal primarily with *descriptive statistics*. This is the use of the subject to describe the results of surveys or experiments. The other main use of statistics is as a tool for *inference*. In this, increasingly important, role the subject is used to extrapolate from small-scale research into larger generalities.

While we shall be drawing some inferences from the work we do, the mathematics and the concepts required for a full understanding of *inferential statistics* cover too much ground to be dealt with satisfactorily in a simple text such as this. My aim here is to provide you with a sound basis for further study of the subject; to provide some health warnings about the activities of Gus and, by moving slowly through the early groundwork, to demystify what is such an important area of our society and our lives.

2 Recording Results

Frequency tables

It is now time for us to look at ways of recording and presenting results. Once we decide to conduct a particular kind of survey and feel fairly confident that we have set it up right – good questions and an unbiased target population – there are available to us several quite different ways of presenting results but generally only one way of recording them. This is in a *frequency table* such as Table 2.1.

Table 2.1. *Analysis of pools results for 9 February*

Result	Number of Matches
Home win	29
Away win	14
No-score draw	4
Score draw	8
Total	55

Underlying much statistical work is the process of counting. We may count how many people have bought the latest record by A La Mode, or how many pop records issued in the month of November have a playing time of 4 to 4.5 minutes, or even how many people prefer Stockhausen to Mozart. In each case we are finding out the number of times which a particular *outcome* has occurred (the *frequency*).

In any investigation there will be a set of outcomes. These are the things which actually happened. They may, for example, be responses to questions in a survey, screws being 1.995 cm long or sales figures for tangerine mousse over a year.

The actual nature of the outcome will depend on the way in which we design the investigation and decide to record the results. Thus in a set of parallel traffic surveys conducted on a stretch of road the passage of a particular car might represent: for survey A, the outcome a red car; for survey B, the outcome a Mercedes Benz; and for survey C it could be the outcome two passengers.

One event can then be recorded in many different ways depending on what we are interested in investigating. Gus likes this potential for manipulation on his part, particularly when trying to 'prove' something by conducting a survey and publishing its results.

Once we have decided how we are going to sort out the responses which we get to a survey, we can record and count how many times each outcome appears.

The definition of *frequency* is the number of times a particular outcome happens in the course of an investigation. For example, you measure a group of twenty people. Seven of them are between 150 and 159cm in height. Seven is the frequency with which the outcome 150–159 cm occurs.

When the survey is over it is usual to record the results in a table like Table 2.2.

Table 2.2.   *Results of a survey of vehicles moving along the T421 between 10.00 and 11.00 a.m., 25 October*

Type of vehicle	Frequency
Light private	46
Light commercial	13
Heavy commercial	16
Articulated	5
Others	3
Total	83

As an aid to preparing such a table the researcher may well make use of a *tally chart* like Table 2.3. Each time a vehicle goes past, the researcher puts a mark or *tally* in the appropriate place. To make it easier to count up at the end, every fifth one is put across like this:

卌   卌   卌   卌   卌   III

rather than this:

If you try to count the two groups above, I'm sure you will agree that the 'fives' method makes it a lot easier to keep track.

Table 2.3

Type of vehicle	Tally						Frequency
Light private	⊦⊦⊦⊦	⊦⊦⊦⊦	⊦⊦⊦⊦	⊦⊦⊦⊦	⊦⊦⊦⊦	⊦⊦⊦⊦	46
	⊦⊦⊦⊦	⊦⊦⊦⊦	⊦⊦⊦⊦	l			
Light commercial	⊦⊦⊦⊦	⊦⊦⊦⊦	lll				13
Heavy commercial	⊦⊦⊦⊦	⊦⊦⊦⊦	⊦⊦⊦⊦	l			16
Articulated	⊦⊦⊦⊦						5
Others	lll						3
Total							83

Grouping data

Often when making a frequency table there are too many possible events to count separately or to record meaningfully in a table.

I can show this best by looking at the responses to a question asked as part of the ten-yearly census conducted by the government of Normalia. The question asks all people in work to report the amount of money that they have earned in the previous year.

Notice that it is highly unlikely that this question will produce an unbiased response! The replies will almost certainly be a low estimate of the true nature of things.

However, having asked the question, the treasury minister, on seeing the results, is not going to be too pleased if they are presented to her as in Table 2.4.

The minister wants to use the results to look for broad patterns in earnings to help her to decide on taxation, and to convince the people that they have never had it so good. With such a large table as 2.4 she will not be able to make out any significant patterns. To help with this problem it is common to group the data together as

Table 2.4

Annual earnings (to the nearest 50 Normalia $)	Frequency
1000	46
1050	84
1100	35
1150	92
1200	46
1250	0
1300	12
1350	27
–	–
–	–
–	–
–	–
24200	1
Total	360000

in Table 2.5. Looking at a table like this one, the minister can, at a glance, say that the majority of people earn between N$7500 and N$17500, and can easily compare this with the results of a previous similar survey.

Table 2.5

Annual earnings (to the nearest 50 Normalia $)	Frequency
0 up to 5000	5000
5000 up to 7500	15000
7500 up to 10000	47000
10000 up to 12500	63000
12500 up to 15000	121000
15000 up to 17500	76000
17500 up to 20000	24000
20000+	9000
Total	360000

A table such as 2.5 is called a *grouped frequency table* because the information in it has been grouped together. Each grouping of data in the left-hand column is called a *class* or *class interval*.

The use of grouped frequency tables is very widespread. It is often necessary, as in this example, to try to spot patterns in what might otherwise be a mass of confusing information. The choice of classes is very important. If we choose too few then the information becomes too simple and underlying trends are in danger of getting lost. (Gus is not above making use of this when it is to his advantage.)

On the other hand, if we choose too many classes we will probably defeat the purpose of the exercise, which is after all to simplify matters. Generally speaking five to ten classes is about the right number to aim for.

Notice that in this example the classes are not all of the same length. (The first goes from 0 to 5000 while the second only goes from 5000 to 7000. The last class has no top limit at all and so we cannot accurately talk about its length.) This variability of length can be both an asset and a liability. Used well, the choice of unequal class intervals will bring out patterns and help to clarify results. In the hands of Gus it becomes a good way to massage statistics, enabling him to highlight his message and to conceal unwelcome inferences (see *How To Lie With Statistics* by Darrell Huff, Pelican, Harmondsworth, 1981).

Having decided on the number of classes we are going to employ, we are now faced with the task of defining them clearly and unambiguously. In the data prepared for the treasury minister it is not clear into which class someone earning N$5000 would be recorded. This table was constructed in an unsophisticated manner. The idea was that the minister, who is noted for a quick tongue but a slow brain, would feel that all possible incomes were covered without any gaps. An alternative, and better, presentation would be that set out in Table 2.6.

It now appears that there is a gap between each class but this is not the case. Because the information was initially recorded correct to the nearest N$50, someone who earned N$4960 (apparently in the gap) would be recorded as earning N$4950 and would, hence, be put into the first class. Similarly a person who earned N$4980 would be recorded as earning N$5000 and would be put into the second class.

This still leaves the problem individual who earns N$4975. This is exactly half way between N$4950 and N$5000. Because this problem of classification comes up very often, the agreed way to deal with it is to put the half-way value into the higher class. So the person earning N$4975 belongs to the N$5000–N$7450 class. We

Table 2.6

Annual earnings (to the nearest 50 Normalia $)	Frequency
0 up to 4950	5000
5000 up to 7450	15000
7500 up to 9950	47000
10000 up to 12450	63000
12500 up to 14950	121000
15000 up to 17450	76000
17500 up to 19950	24000
20000+	9000
Total	360000

say that N$4975 is the *class boundary* between the first and second classes. Similarly N$7475 is the boundary between the second and third classes, and so on. By agreement the boundary value belongs to the upper class.

Table 2.7

Annual earnings (Normalia $)	Frequency
0 up to 4975	5000
4975 up to 7475	15000
7475 up to 9975	47000
9975 up to 12475	63000
12475 up to 14975	121000
14975 up to 17475	76000
17475 up to 19975	24000
19975+	9000
Total	360000

This means that we can present the frequency table as in Table 2.7. In this case we can leave out the information that the data was collected to the nearest N$50 because that is taken care of in the way we have written the classes. Table 2.8 shows a few alternatives for recording these classes which you will come across in various places.

To be well written it must be immediately clear to the reader into which class any particular value will be placed. This becomes of

Table 2.8

Normalia $ (nearest 50)	or	Normalia $ (nearest 50)
0-		less than 5000
5000-		5000 but less than 7500
7500-		7500 but less than 10000
10000-		10000 but less than 12500
12500-		12500 but less than 15000
15000-		15000 but less than 17500
17500-		17500 but less than 20000
20000-		more than 20000

particular importance when we start to calculate averages and to draw diagrams.

Exercise 2.1

1. After a fishing competition all of the fish caught were weighed and the results recorded. All weights are in kg (correct to the nearest 100g)
   2.2, 1.0, 1.9, 1.7, 1.3. 1.2, 2.1, 1.3, 1.1, 1.6, 1.7, 1.5, 1.6, 1.6, 1.9, 1.8, 1.1, 1.3, 1.4, 1.7, 1.2, 1.9, 1.9, 1.8, 2.0, 2.1, 2.1, 1.2, 2.0, 1.4, 2.0, 1.3, 1.6, 1.6, 1.9, 1.7, 1.6, 1.6, 1.2, 1.6, 1.5, 1.7, 1.2, 1.1, 1.6, 1.7, 1.5.
   Convert this information into a frequency table.

2. As part of a research programme into sources of protein, workers at the Normalia Institute of Agriculture are investigating different strains of soya bean with a view to improving its quality. Of sixty-four new varieties grown, these are the average yield of energy expressed as Calories per 100g of beans (correct to 1 decimal place).
   196.5, 180.4, 178.1, 188.2, 154.9, 185.9, 188.9, 161.4, 159.6, 194.2, 171.3, 202.4, 174.1, 186.2, 166.4, 172.8, 197.1, 188.1, 187.2, 161.0, 162.3, 174.5, 165.0, 183.5, 162.7, 170.9, 167.9, 190.0, 157.7, 198.5, 162.1, 178.7, 192.5, 187.4, 188.7, 170.6, 166.5, 189.1, 152.1, 180.0, 167.9, 169.0, 199.7, 163.2, 204.6, 184.1, 186.0, 204.6, 180.2, 158.0, 182.5, 174.0, 197.6, 186.5, 194.4, 170.8, 169.1, 187.8, 172.2, 183.8, 194.1, 172.7, 166.4, 158.5.

Convert this information into a grouped frequency table taking class-intervals of length 10.0.

3. For each of the following extracts from grouped frequency tables state, with reasons, the class boundary between the two classes shown.

(a) Length of screws		
(cm to nearest mm)	1.5–1.9	2.0–2.4
Frequency	47	83
(b) Age in years of pupils	7–9	10–14
Frequency	26	35
(c) Length of long jump		
(cm to the nearest 10 cm)	340–350	360–370
Frequency	12	15

Presenting results

When we have finished recording all the results of an investigation, it is then necessary to present them in a fashion which makes them easy to understand and to interpret. In some cases leaving them in a frequency table may be adequate but often some form of a picture is used to represent the results. There are several different types of pictures and each has its advantages, its drawbacks and its dangers.

Bar charts

These are a fairly obvious extension of the idea of a tally chart. A series of bars, either horizontal or vertical, is drawn. The length of each bar represents the frequency of the event for which it stands. On simple bar charts like these all the bars should be the same width. So the results of Table 2.2 about a traffic survey might be presented as in Figure 2.1.

There are four particularly important features of any bar chart:

(a) The title. Without this to explain what it shows, the chart's value is considerably diminished. This is true of all tables

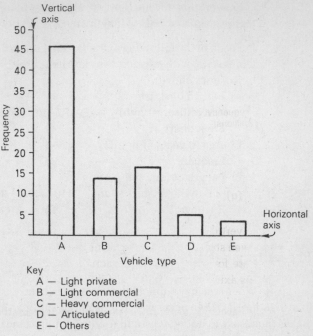

Key
A — Light private
B — Light commercial
C — Heavy commercial
D — Articulated
E — Others

Figure 2.1.   *Bar chart showing vehicles recorded in one hour of survey*

and charts in statistics. It is very frustrating to meet a beautifully presented chart and to be unable to find out what it is about. Newspapers and magazines are often very bad at this. A good chart should make sense on its own without the reader having to wade through accompanying text trying to find out what it is supposed to show.

(b) The label on the horizontal scale (axis). This shows what the bars stand for. Without it you can end up with a very pretty chart that conveys no information.

(c) The word 'Frequency' on the vertical axis. This is all too often left off and is a source of confusion. In this example I could have used the words 'Numbers of Vehicles' instead. If you have used any form of scaling here it is necessary to explain it. By this I mean that if our survey involves very large frequencies (millions, say) then it is easier to use a shortened form for labelling the axis and to explain the abbreviation beside. So, Figure 2.2(a) is preferable to Figure 2.2(b).

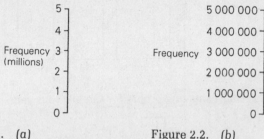

Figure 2.2. *(a)*   Figure 2.2. *(b)*

(d) The vertical (frequency) scale is unbroken. It starts at 0 and goes up in even steps the whole way. Gus revels in the opportunities available here for deceit and malpractice. By interfering with the scale on this axis Gus can make insignificant changes seem vastly important.

Take the case of the annual report of the Imaginative Designs Company to its shareholders. Part of this report is a chart showing profits over a four-year period (Figure 2.3).

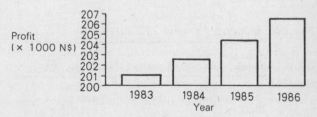

Figure 2.3. *Chart showing profits for the last four years*

This is the chart prepared for the directors by our friend Gus and it shows that the company's performance has improved dramatically over the four-year period. Or does it? A more accurate diagram would be like Figure 2.4. It should now be more obvious to the shareholders that, while the company is still making profits, these have not increased significantly over the last four years.

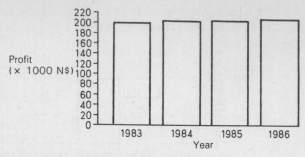

Figure 2.4.   *Chart showing profits for the last four years*

Grouped data

Let's go back to a slightly amended version of Table 2.5 about the earnings in Normalia (Table 2.9). If we are using grouped data as in Table 2.5 then our chart should look like Figure 2.5. I cheated and amended the original table to make drawing the chart easier. You will notice that in my new table all the classes have the same length (N$2500). This makes life simpler. If the classes have different lengths then we are presented with a more difficult task. I will deal with that later.

Table 2.9   *(Amended version of Table 2.5)*

Annual earnings (to the nearest 50 Normalia $)	Frequency (nearest 1000)
5000– 7500	15000
7500–10000	47000
10000–12500	63000
12500–15000	121000
15000–17500	76000
17500–20000	24000
Total	346000

The notable feature of Figure 2.5 is the marking on the horizontal axis. I have chosen to mark the centres of the class intervals on the scale and to make these the centres of the bars. It is possible to draw

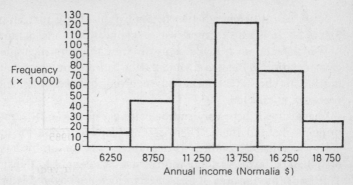

Figure 2.5. *Bar chart showing annual earnings for citizens of Normalia*

a bar chart like Figure 2.1 to represent this information but in this case the horizontal axis would be labelled as on Figure 2.6.

Notice that here I have labelled the axis using the class intervals. I should, to be accurate, use the class boundaries instead so that the first bar would represent $4975–7575, as previously discussed. The same would apply to all the other bars. In this case I feel that this is probably taking accuracy too far. Because precision in this area can be confusing and hence counterproductive, we must decide, when drawing a diagram such as this, whether clarity or pedantry is more important.

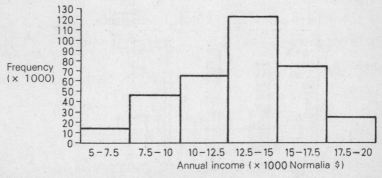

Figure 2.6. *Bar chart showing annual earnings for citizens of Normalia*

There is a difference between this bar chart and that drawn in Figure 2.1. In Figure 2.1 there was no connection between the types of vehicles and so it made sense to draw the bars separately for greater clarity. In Figures 2.5 and 2.6 the bars are drawn so that they touch. This shows that the classes involved abut each other and are, probably, connected.

I have shown here very rudimentary bar charts to illustrate the principle behind them. There is considerable scope for artistic enhancement. One common device is to replace the bars by drawings of cylinders, boxes or some representation relevant to the survey. This leads into the area of pictograms which are covered later.

Another type of bar chart displays two or more sets of data on the same diagram, by placing bars either beside each other as in Figure 2.7(a) or on top of each other as in Figure 2.7(b).

The second type of diagram here is not a very good one since it requires rather a lot of work to decipher. This is due to the first bars being of varying heights. A better use of a diagram based on this principle is the *percentage composite bar chart* (Figure 2.8). In this case it is easy to compare the relative importance of items within the overall total. The problem now is that we have no idea of the relative values of the imports. Generally this type of comparison can be achieved better on a *pie chart*, which we shall meet soon.

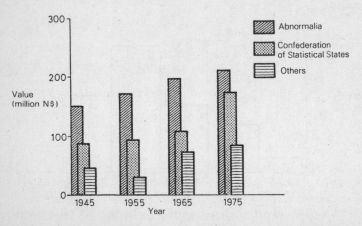

Figure 2.7.   (a) Comparison of import sources to Normalia

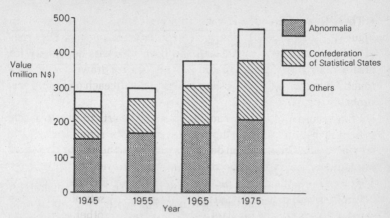

Figure 2.7. (b) Comparison of import sources to Normalia

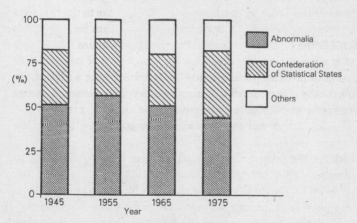

Figure 2.8. Comparison of import sources to Normalia

Exercise 2.2

1/2. Draw bar charts for the data from Exercise 2.1.
3. In a research project into the efficacy of various fertilizers, a researcher grew two batches of potatoes, one in fertilized and one in unfertilized soil. Her results are shown in Table 2.10.

Table 2.10

Weight of potatoes (nearest g)	Frequency with fertilizer	Frequency without fertilizer
0 –	17	23
50 –	30	37
100 –	54	72
150 –	63	85
200 –	84	64
250 –	49	27
300 –	12	9

Draw two bar charts showing these results separately. Then draw a combined bar chart with both sets of data beside each other. Which method allows for easier comparison? Do you think that the fertilizer had any effect?

Histograms

As promised, I shall now deal with charts for grouped data where the classes have differing lengths. This section is going to be rather complicated but is worth reading as it provides the basis for a lot of more advanced work in presenting and interpreting statistics.

I need, for the sake of clarity, and because it is widely used, to introduce now another technical term – *variable*. This is really just another word for the thing which a survey is measuring or counting.

As we have seen, in any survey there are various possible outcomes. Each time that we record the result of an experiment or the answer to a survey question this may *vary* among a set of possible results or answers. So the result is usually referred to as the *variable* in the experiment or survey.

If we were to conduct a survey into the ages of a group of 100 people, the variable would be their ages because this would vary from person to person. In another survey of road usage the variable might be the type of vehicle going past the survey position or it could be the colour of the vehicle or the number of passengers inside it depending on the design of the survey.

Variables are split into two types, *qualitative* and *quantitative*. The difference is quite straightforward. Qualitative variables are ones such as taste, colour, attractiveness which are not easily expressed as a number. Quantitative variables, on the other hand, are numerical in nature and are, as a consequence, much more easily processed by statisticians using traditional arithmetic and algebra. Having said that, the branch of Statistics (non-parametric Statistics) which is able to handle both types of variable is becoming increasingly popular and important. We shall touch on this a little in a later chapter.

Quantitative variables are usually in turn divided into two groups – *continuous* and *discrete*. Discrete does not mean that they are careful about what they say but rather that the variable will only take certain values and will never fall between those values. A continuous variable on the other hand can take any value at all. There will be no breaks in its set of possible values.

In the examples which I used above, people's ages would be a continuous variable since a person responding to the survey could be any age at all between 0 and 120 years. (We would not really expect them to be over 120.) On the other hand, if the survey into road usage recorded the number of passengers in each vehicle, then this would be a discrete variable since you can only have whole numbers of passengers. Two-thirds of a passenger would be rather messy! In general, discrete variables tend to take whole-number values although there are exceptions. Shoe sizes for instance are discrete despite there being sizes such as 8½.

We are now in a position to deal with continuous variables and, more particularly, to draw diagrams to represent them. When we dealt with events recorded as classes we were, in fact, dealing with continuous variables. When I drew a bar chart for this kind of information (Figure 2.5), if you remember, I cheated and changed the original data to make it easier to draw the chart. I now want to make up for that trick and to try to show how to draw a good diagram for data which is not so well behaved and which comes in classes of unequal length. So first of all some not quite so orderly data . . .

As part of a survey of national fitness the Ministry of Health in Normalia selected fifty adults at random. One of the first things which they did was to weigh them and to record the information

(Table 2.11). It all seems fairly innocuous so far, but look at the bar chart for this (Figure 2.9).

Table 2.11.   *Weights of a sample of fifty adults*

Weight (nearest kg)	Frequency
30–44	5
45–54	7
55–59	10
60–64	9
65–69	7
70–79	6
80–99	6
Total	50

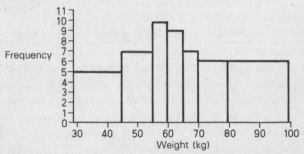

Figure 2.9.   *Bar chart showing the weights of fifty adults in Norm-alia*

It is, I hope, obvious that something has gone wrong here! The diagram seems to say that the majority of the people are either very heavy or very light. This is not the case, but it seems so because our eyes and our brains respond to the areas of the bars even though we know that it should be the heights.

It is to avoid this kind of illusion, and also for more advanced reasons which will make later mathematics easier, that we usually draw a diagram called a *histogram* instead. Here frequencies are represented by *areas* and *not* by heights. Our diagram now looks like Figure 2.10. This is, I am sure you will agree, an obvious improvement. The calculation needed before drawing the histogram

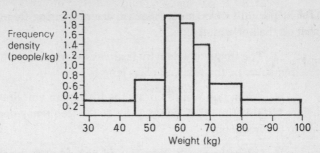

Figure 2.10. *Histogram showing the weights of fifty adults in Normalia*

is not too bad (Table 2.12). There are a few things to watch out for here.

Table 2.12. *Weights of a sample of fifty adults*

Weight (nearest kg)	Frequency	Class length	Frequency density
30–44	5	15	5/15 = 0.33
45–54	7	10	7/10 = 0.70
55–59	10	5	10/5 = 2.00
60–64	9	5	9/5 – 1.80
65–69	7	5	7/5 – 1.40
70–79	6	10	6/10 = 0.60
80–99	6	20	6/20 = 0.30
Total	50		

1. It is tempting to work out the class length incorrectly. With the data presented in this form – to the nearest kg – the class length can mistakenly be calculated to be one unit too small. In the case of the 65–69kg class the class length is *not* 4kg (69 − 65 = 4). Since the figures are given correct to the nearest whole kg, the lower limit of the class is 64.5kg while the upper limit is 69.5. The class length is, therefore, 69.5 − 64.5 = 5kg. This applies throughout the table.

2. Frequency has been replaced by frequency 'density'. The analogy here is with the density of wood which is measured in weight per unit of volume. Here we are measuring frequency per unit on the horizontal axis.

3. The frequency density is always worked out by dividing the frequency by the relevant class length.

4. When you want to draw the histogram you draw bars as before but this time they go up as far as the *frequency density* and not to the frequency.

5. If you have a histogram and you want to work out the frequency represented by a bar you just multiply the height of the bar by its width (i.e. find its area).

Exercise 2.3

1. The figures in Table 2.13 record the class sizes for all of the primary schools in Normalia. Draw a histogram to represent this information.

Table 2.13

Classes of size	Pupils in this group (%)
0–15	2.8
16–20	4.2
21–25	8.3
26–30	14.7
31–35	31.9
36–45	29.6
46–60	8.5
Total	100.0

2. The Modal Beer Company produces several different kinds of beers. As part of their quality-control process they measure the specific gravity of each batch when fermentation is complete. Figure 2.11 shows the results for 100 batches of Katz Bitter represented on a histogram. Convert this diagram into a frequency table. What percentage of the batches had a specific gravity of less than 996.5?

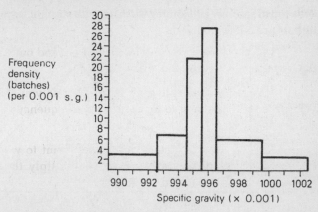

Figure 2.11.   *Histogram showing final s.g. for 100 batches of Katz Bitter*

3. As part of its information service, the Meteorological Office of Normalia published Table 2.14 concerning rainfall over a year. Construct a histogram to illustrate this data.

Table 2.14

Depth (inches to 1 d.p.)	Number of days
0–0.5	104
0.6–1.0	123
1.1–1.5	84
1.6–2.5	23
2.6–3.5	9
3.6–5.0	22
Total	365

Frequency polygons

Often, to make patterns clearer, instead of drawing a bar chart or a histogram we draw another type of diagram. One such is the *frequency polygon*. This is made by putting a dot in the centre of the top of each bar and joining them up with straight lines. Usually the bars are not even drawn.

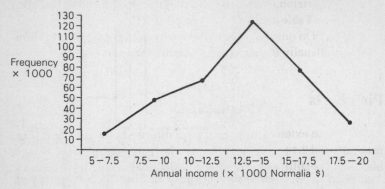

Figure 2.12.   *Chart showing annual earnings for citizens of Normalia*

Very often more than one frequency polygon is drawn on the same diagram in order to make easier the comparison of two or more sets of data. For example the income of the citizens of Normalia for this year and for last.

Exercise 2.4

1. In an attempt to make their version of rugby more attractive to spectators, the Normalia Rugby Union amended

Table 2.15

Number of tries per match	Season	
	Before change	After change
0	184	142
1	243	208
2	152	249
3	74	89
4	59	48
5	27	16
6	12	6
7	6	1
8	3	1
Total	760	760

their rules. This increased the points scored for a try from four to six. Monitoring of the number of tries scored in Division One of their Merit Table produced the statistics in Table 2.15.

On one diagram, draw two frequency polygons to show visually the success or failure of the rule change.

Pie charts

An extensively used type of diagram is the *pie chart* (Figures 2.13 and 2.14). This gets its name because of its obvious resemblance to the cutting up of a pie. It is slightly more difficult to draw

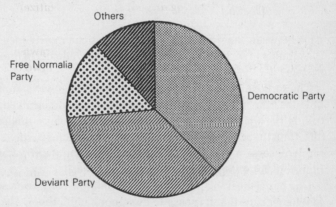

Total votes cast = 1 635 247

Figure 2.13. *Pie chart showing voting patterns in Normalia General Election*

than a bar chart but it has the advantage that it seems to make it easier for the human eye to absorb information.

There is little to say about interpreting pie charts as they are so self-explanatory. The bigger the slice, the greater the thing it stands for. In the two examples we can see that the party which got the biggest share of the votes, the Democratic Party, also won the majority of the seats. The two diagrams show up, however, the differ-

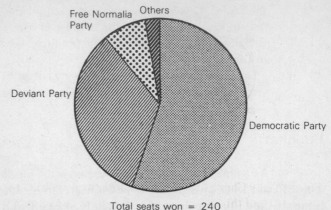

Total seats won = 240

Figure 2.14.   *Pie chart showing seats won at Normalia General Election*

ence between the proportions of the votes each party received and the number of seats these votes earned them. A defect of the electoral system?

The sort of numbers being considered are not always immediately obvious from a pie chart. This information should be given, as here, beside the diagram. It is then possible to measure angles on the chart and thence to calculate the actual frequencies involved. This is rarely necessary nor is it likely to be very accurate. (Protractors are notorious for giving people the wrong answer!)

Gus has a sneaking partiality for pie charts. They seem so simple that he reckons that it is easy to fool people with them. The Unlimited Expansion Company Limited (manufacturers and installers of cavity wall insulation) call on Gus to prepare some advertising material for them based on the fact that their annual turnover has doubled in a twelve-month period. Gus comes up with two pie charts (Figure 2.15).

When queried about the accuracy of these diagrams, Gus can reply with a straight face that, if you measure them, the second circle's diameter is twice that of the first and that he would have given the numbers involved but he didn't want to confuse matters! The assertion about the diameters is certainly true, but it means that the second circle is actually four times bigger than the first . . .

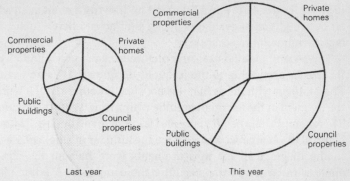

Figure 2.15.   *Charts to show the performance of UEC Ltd last year and this year*

Pictograms

These are probably the most widely used and abused methods for showing statistical information. The principle is very simple and is based on the idea that instead of using bars or sectors of a circle to represent frequencies we draw a picture which is in some way associated with the survey.

The size of the picture is supposedly in proportion to the frequency. There are two different types of pictogram used. The first, Figure 2.16, is really just a glorified bar chart.

This sort of illustration is pretty but is usually rather tedious to draw. Except in situations like television or newspaper presentation

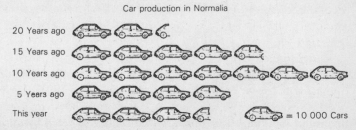

Figure 2.16.   *Car production in Normalia*

of statistics the extra effort involved in drawing a chart such as this far outweighs any advantages over a simple bar chart which can, in fact, be much more accurate.

The second method uses one picture for each event and changes its size in proportion to the frequency. Figure 2.17 is an example. The problem with this is that it is not clear whether it is the height of the diagram or the apparent volume which represents the frequency. This example tells lies visually and is much loved by Gus. The actual effect of the campaign was to halve the number of smokers. Because in the diagram all the measurements were halved, the cigarette appears to be 1/8 of the size after the campaign. A fairer picture could be one where only the length has changed. Alternatively we could draw a scaled cigarette which was half the size. This involves dividing *all* of the measurements by the cube root of two (1.26 to two decimal places). The calculations involved in the former model are so much easier that it is the one I recommend to you. Figures 2.18 and 2.19 show the two versions.

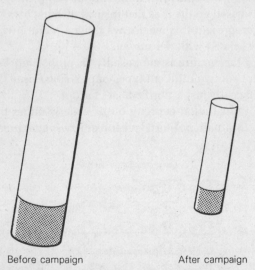

Before campaign   After campaign

Figure 2.17.   *Number of smokers in a school affected by an anti-smoking campaign*

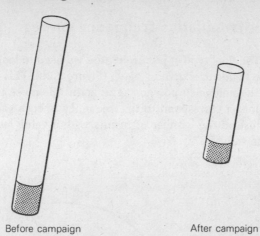

Before campaign   After campaign

Figure 2.18.   *Number of smokers in a school affected by an anti-smoking campaign*

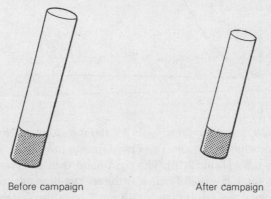

Before campaign   After campaign

Figure 2.19.   *Number of smokers in a school affected by an anti-smoking campaign*

To draw a reasonable pictogram of this form requires mathematical knowledge of areas and proportions beyond the scope of this book. If you want more information then look up in an ordinary Maths textbook the topic called 'ratio and proportion' – and good luck!

B.S.—3

Ogives or cumulative-frequency curves

The final type of picture which we need to look at is the *cumulative frequency curve* or *ogive* (Figure 2.20). This is closely related to the frequency polygon. The main difference is that the points no longer correspond to the frequency of each variable but represent instead the number of results with a value *less than* or *equal to* the variable.

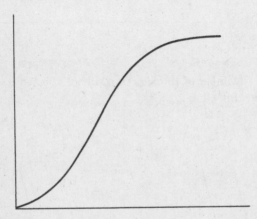

Figure 2.20.

I shall use the figures from Table 2.7 about earnings in Normalia to demonstrate the technique. The first step is to prepare a cumulative frequency table (Table 2.16). The right-hand column is really just a running total. It is a useful check at the end that the total of column 2 should equal the last number in column 3. Note also that the 251 000 in column 3 means that 251 000 people had an income of $14 975 or less.

To translate this table into a graph (Figure 2.21) is quite straight-forward. Notice that unlike the bar chart or the frequency polygon, I have here marked the *ends* of the class intervals on the horizontal axis. This is because the graph represents values *up to and including that amount*.

Also, although for the reasons which we have already met, the

Table 2.16.   *Annual earnings for citizens of Normalia*

Annual earnings (Normalia $)	Frequency (nearest 1000)	Cumulative frequency
0– 4975	5000	5000
4975– 7475	15000	20000
7475– 9975	47000	67000
9975–12475	63000	130000
12475–14975	121000	251000
14975–17475	76000	327000
17475–19975	24000	351000
19975+	9000	360000
Total	360000	

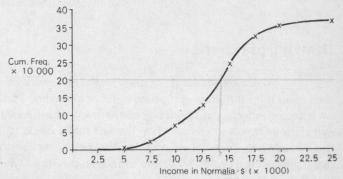

Figure 2.21.   *Cumulative-frequency curve showing annual income in Normalia*

class boundaries are 4975 etc., it is easier and not too misleading to round the figures off when plotting. So 4975 is plotted at 5000 and so on.

Finally, the last entry on the first cumulative frequency table was for people who earned $19975+. Since there is no upper limit on this figure I have chosen, arbitrarily, to make $25000 the upper limit on the assumption that not very many people will earn more than this. My assumption may quite easily be wrong, but the numbers involved are small enough not to worry me at this stage.

This graph provides us with the first opportunity to do more than just describe the statistics we have recorded. We can now start to use them to gain other information. Looking at the curve, it is reasonable to draw in the dotted lines shown, and to use them to estimate that about half the population earn less than $13 500 and about half earn more than that amount. We shall return to this later.

Note: This kind of graph is called a cumulative-frequency curve and must be drawn as such. Resist the temptation to join the points with straight lines.

Exercise 2.5

1/2/3. Represent the data from Exercise 2.3 as cumulative-frequency curves.

Drawing pie charts

To draw a pie chart we must first of all draw a circle and then chop it up into the appropriate number of sectors. The size of each sector reflects the frequency of the event it represents. The calculations involved are based on the fact that a circle is usually divided into 360 degrees. I must assume that you know how to measure angles using a protractor. If you cannot do this, then I am afraid that this bit is not for you.

The stages are:

(1) Find the total frequency for the survey.
(2) Divide 360 by this number to get the number of degrees represented by one occurrence of each event.
(3) For each event multiply the answer to (2) by the relevant frequency.
(4) Draw a line from the centre of the circle to the outside and measure each sector in turn using the angles calculated in (3).
(5) Label the diagram as shown. There is no need to mark on it the size of the angles, which I do in my example merely to bring out the above points.

As my example I shall take the results of a survey of sixty people who were asked where they spent their last holiday (Table 2.17).

Table 2.17

Place	Frequency
Home	12
UK	18
Europe	20
N. America	7
Elsewhere	3
Total	60

(1) Total frequency = 60

(2) 360° divided by 60 = 6°

Table 2.18

Place	Frequency	Angle (°)
Home	12	12 × 6 = 72
UK	18	18 × 6 = 108
Europe	20	20 × 6 = 120
N. America	7	7 × 6 = 42
Elsewhere	3	3 × 6 = 18
Total	60	360

(3) It is always a good idea, as I have done in Table 2.18, to add up your calculated angles *before* drawing the pie chart. This can save wasting time if you have made a simple arithmetic error.

(4) and (5) Figure 2.22

In my example I chose a survey of sixty people to make the arithmetic easy. Usually things don't work out so nicely. With the help of a calculator, however, this is no real problem. The only difficulty is that the answer to (2) can turn out to be a rather long decimal.

For example, if the survey involved a total of seventy people then the angle for each person would be $360°/70 = 5.1428571°$. It is obviously impossible to measure an angle to this degree of accuracy, so it is better to round it off to the nearest degree, making each person's sector 5°.

The problem with this is that $70 \times 5 = 350$ so after drawing the

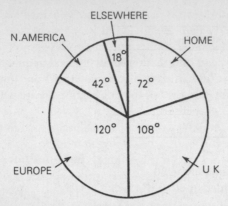

Figure 2.22. *Pie chart showing holiday destinations for sixty people*

chart there will be a sector left over. The solution to this is to save the rounding until after step (3). To make life easier, store the answer to (2) in your calculator's memory and recall it for each multiplication.

I will illustrate this by using the same survey but with slightly amended results in Table 2.19.

Table 2.19

Place	Frequency	Angle(°)	
Home	11	$11 \times 6.5454545 = 72$	(72)
UK	16	$16 \times 6.5454545 = 104.727272$	(105)
Europe	18	$18 \times 6.5454545 = 117.818181$	(118)
N. America	8	$8 \times 6.5454545 = 52.363636$	(52)
Elsewhere	2	$2 \times 6.5454545 = 13.090909$	(13)
Total	55	360	

$(360/55 = 6.5454545)$

You may find that the total angle works out to be more, or less, than 360. If the difference is only one or two degrees, then do not worry as you probably will not be measuring your angles all that accurately, and the error is only caused by the rounding process.

Pie charts can become cluttered if they are used to present too much information. I would recommend that you do not use this type of chart if you have more than eight different events to represent.

Finally a word about the size of pie charts. It is possible to incorporate into a pie chart information about the total frequency or value represented by it. In this case it is the area and not the radius or diameter which should convey the information because this is what the brain, via the eye, absorbs.

In general the total frequency should be in proportion to the area of the circle. This in turn means that it should be in proportion to the square of the radius, i.e. if we double the radius in a pie chart it should represent four times the frequency (2 × 2). If we want to show the frequency being doubled then it is necessary to multiply the radius by $\sqrt{2}$. A typical calculation would look like this.

> For pie chart 1 total frequency = 100.
> For pie chart 2 total frequency = 120.

> Radius of pie chart 1 = 5cm
> Radius of pie chart 2 = 5 × $\sqrt{(120/100)}$
> $\qquad\qquad\qquad\quad$ = 5 × 1.0954
> $\qquad\qquad\qquad\quad$ = 5.4770
> $\qquad\qquad\qquad\quad$ = 5.5cm (1 d.p.)

Exercise 2.6

1. As part of a survey of road usage the information in Table 2.20 was recorded about vehicles parked along a street. Draw a pie chart to represent this information.

Table 2.20

Place of origin	Frequency
UK	23
W. Europe	17
E. Europe	2
Japan	15
Others	3
Total	60

2. Yahboo Magazine asked its readership to say how they spent the greater part of the hour from 7 p.m. to 8 p.m. on a particular Tuesday. Their replies are shown in Table 2.21. Show this data on a suitable pie chart.

Table 2.21

Occupation	% of Respondents
Watching TV	46
Doing homework	15
Eating	4
Reading	6
Listening to music	22
Other	7
Total	100

3. Having adopted a neutral stance in regional affairs Normalia has been able, over a period of five years, to reduce its defence spending. A breakdown of how this has been achieved is shown in Table 2.22. Draw two appropriately sized pie charts to show this information.

Table 2.22.   Spending ($\times$ N$10 000 000)

Area	5 years ago	Present
Navy	47	25
Air Force	32	20
Army	54	18
Intelligence	7	7
Total	140	70

3 Patterns in Surveys

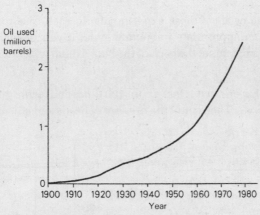

Figure 3.1. *Fuel consumption in Abnormalia*

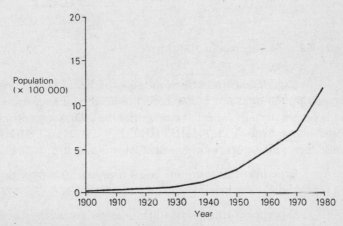

Figure 3.2. *Population of Abnormalia*

If we did a lot of surveys and statistical investigations, certain themes or patterns of results would emerge. These characteristic *frequency distributions* are worth looking at a bit more closely because they provide the clue to links between seemingly unconnected situations and sets of data. Here are a few experiments for you to try. They are a little tedious but worth doing because they illustrate some of the more common patterns we come across in statistics.

Try some or all of these experiments. In each case record your results in an appropriate frequency table. If you prefer, there are computer programs at the back of the book which will simulate them for you.

Experiment 1: toss a coin 100 times and record your results THHTHT, etc. Then count the number of heads and tails that result.

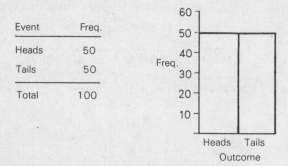

Event	Freq.
Heads	50
Tails	50
Total	100

Figure 3.3.   *Tossing a coin 100 times*

Experiment 2: take the results of experiment 1 and, instead of counting the total number, record the *lengths* of sequences of heads. Ignore any tails except as markers for the beginnings and ends of sequences. Thus TTHTHHHTHHTHTTHTTHTTTHHHHT would represent sequences of heads of lengths 1, 3, 2, 1, 1, 1, 4.

Experiment 3: toss four coins together ninety-six times and record how many heads show up on each throw.

Experiment 4: toss a die sixty times and record the number on top.

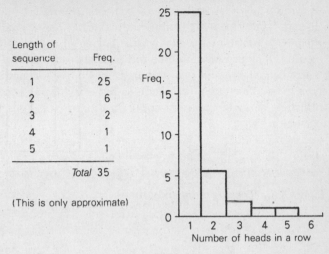

Length of sequence	Freq.
1	25
2	6
3	2
4	1
5	1
Total	35

(This is only approximate)

Figure 3.4. *Lengths of sequences of heads when tossing a coin*

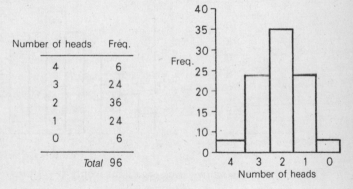

Number of heads	Freq.
4	6
3	24
2	36
1	24
0	6
Total	96

Figure 3.5. *Tossing four coins ninety-six times*

Experiment 5: toss two dice together seventy-two times and record the total of the two numbers showing.

Draw bar charts showing the results of each experiment. You now have diagrams showing actual *frequency distributions*. This means the ways in which the frequencies are distributed or spread among the different possible events. Later on in the book you will learn

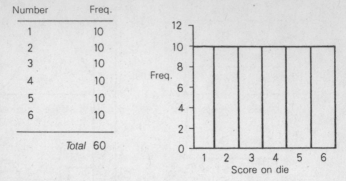

Number	Freq.
1	10
2	10
3	10
4	10
5	10
6	10
Total	60

Figure 3.6.   *Tossing a die sixty times*

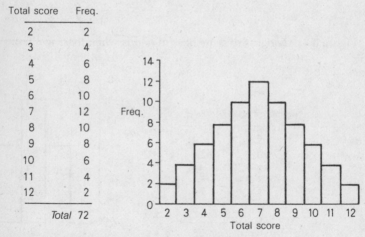

Total score	Freq.
2	2
3	4
4	6
5	8
6	10
7	12
8	10
9	8
10	6
11	4
12	2
Total	72

Figure 3.7.   *Tossing two dice together seventy-two times*

how to predict some of these frequency distributions without doing the experiments.

Figures 3.3 to 3.7 show the theoretical results to the experiments. I would be very surprised if any of your results match these exactly but they should be reasonably close.

It turns out that certain types of frequency distribution come up again and again in different situations. The sketches in Figure 3.8 show the general shapes associated with them.

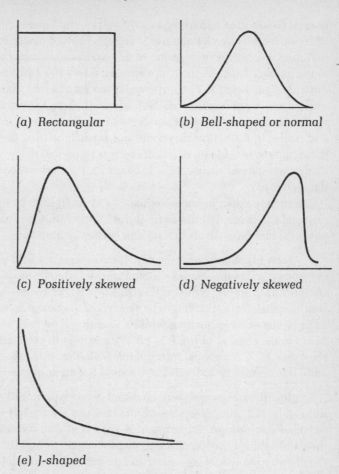

(a) Rectangular

(b) Bell-shaped or normal

(c) Positively skewed

(d) Negatively skewed

(e) J-shaped

Figure 3.8.

(a) gets its name for obvious reasons and as you can see describes the theoretical results of experiments 1 and 3.

(b) is by far the most common type of distribution you will find when doing statistical surveys. It is usually called the *Normal Distribution*. One reason for this is that it is the distribution which you *normally* come across. The mathematics of the distribution is rather complicated but if you want a simple description of the normal distribution and how to use it, you can find one in *Statistics Without Tears* by Derek Rowntree (Penguin, Harmondsworth, 1981).

Almost any survey or experiment will yield data which conforms to the normal distribution. Supposing we take 1000 adult men and measure their heights. Then, if we put the results into classes, we will get a histogram which will fit fairly well under a normal curve. If we make the class intervals smaller and smaller, we will get nearer and nearer to fitting the curve without actually getting there. Our *actual* survey is made up of lots of *discrete* (separate) data while the *theoretical* normal distribution is based on the idea of *continuous* data (no gaps).

Fortunately the differences are usually so small as to be insignificant and the theory behind normal distribution provides us with a powerful tool for analysis of data and inference from it.

(c) & (d). Sometimes the distribution does not turn out to be as symmetrical as the normal distribution and something causes the central hump to be pushed either to the right or to the left. The resulting distribution is said to be *skewed*. If the hump is to the left as in (c) the skew is *positive* while a *negative* skew will move the hump to the right as in (d). It is possible to quantify the amount of skew (see K. A. Yeomans, *Introducing Statistics*, pp. 114–17, Penguin, Harmondsworth, 1968) but we shall not attempt that here.

(e). If the skew gets very large and positive (or negative) then, ultimately, the hump vanishes off the end and we are left with the *J-shaped distribution*. Experiment 3 produced a distribution like this, and again if the experiment was repeated enough times then the distribution would get nearer and nearer to the smooth J-shaped curve.

Exercise 3.   For each of the following try to decide which frequency distribution is going to be the best description of what is happening.

   1. A card shop records the total number of birthday cards sold each month for a calendar year.

   2. The same card shop records the number of Christmas cards sold per month for each month throughout the year.

3. In the annual 16+ examinations in Normalia 10000 candidates take the Statistics examination. They are awarded a mark out of 100 and the number of people gaining each mark is recorded.

4. To fight a virulent strain of bacteria, medical researchers at the University of Normalia are testing a new drug. They start with a known number of the bacteria to which they add the drug. They record at the end of each hour the number of bacteria remaining and find that in each hour the drug kills 25 per cent of the bacteria still alive.

5. As part of a traffic survey in Normalia the Department of Transport employs two students to count the number of vehicles travelling down the C47, a minor road. The students find that about a hundred cars an hour go past them. After a while, to make the job a little more interesting they start to bet on how many cars will pass in the next minute. They record the results over a one-hour period like this: 0, 1, 1, 2, 0, 0, 1, 4, . . . At the end of the hour they sort these out and count the number of minutes with 0 cars, 1 car, etc. They then draw a bar chart showing the results.

4 Representative Values

A picture such as those we have met in the last two chapters gives us a good idea of the nature of a distribution. However, if we want to go any further, and to use the results of our surveys to draw conclusions about the total population or to make predictions about its behaviour, then it is necessary to start to process the figures themselves.

The first step in this direction is to simplify the large collection of data into one or two pieces of information called *representative values*. These are usually numbers but may occasionally be otherwise. It is only when they are in a numerical form that they lend themselves to further analysis.

The centre – averages

The first statistic used as a representative value is some measure of the centre of the distribution. This is usually called the *average*. Thus, the average man in the street is a mythical person who in some way is representative of the total population and who is not too extreme in his views.

The average is useful because it provides some idea of the nature of the data being studied. There are three commonly used averages, although one of them has three versions. Each has its advantages and its disadvantages.

The mode or modal class

This is the event which comes up most often. The word comes from the French word for fashion and so the *mode* or *modal value* is the most fashionable or popular one. If the mode is represented by a class interval rather than a single value then this is referred to as the *modal class.*

Every week in Normalia there is a survey done of how many records are sold. The record which sells the most copies is the *mode* for the distribution of record sales. Of course it is usually referred to as the Number One record and the frequency distribution is more commonly called the Record Charts or the Top 100.

If there is no single event which is more popular than the others then the distribution is said to have no mode. Occasionally, if two events share the highest frequency, then the distribution is called *bimodal.*

This average is the only one of the three which can deal with non-numerical data. In other words the mode may be a colour or a taste or anything for which people might express a preference.

Exercise 4.1

1. As part of an experiment in telepathy fifty people were asked to think of a number between 1 and 10. These were their replies.

   10, 8, 8, 10, 3, 3, 8, 10, 2, 8, 10, 8, 4, 5, 6, 7, 10, 2, 10, 6, 6, 8, 2, 9, 6, 6, 3, 3, 8, 4, 6, 4, 6, 7, 4, 3, 2, 3, 9, 3, 6, 8, 8, 2, 9, 8, 4, 8, 4, 8.

   Convert this data into a frequency table and use it to find the modal number.

2. To promote the sales of hats the Normal Attire Company went out into the streets and measured lots of people's heads. Here are some of their results (in cm).

   56, 54, 51, 57, 51, 57, 50, 59, 54, 56, 56, 54, 56, 53, 57, 55, 55, 58, 58, 54, 53, 54, 57, 56, 58, 55, 54, 59, 57, 51, 51, 55, 59, 53, 52, 55, 54, 58, 56, 54.

   By constructing a frequency table, find the mode for this distribution.

3. The Sneezeless Snuff Company Ltd, in an attempt to push the idea that their product lived up to its name, carried out tests with two groups of thirty people. One group was given Sneezeless Snuff and the other received a leading

competitive brand. Their sneezes after one pinch of snuff in each nostril were recorded.

Group A: 3, 5, 3, 5, 3, 2, 3, 1, 6, 2, 1, 1, 5, 2, 3, 1, 4, 1, 6, 5, 3, 1, 6, 6, 2, 5, 5, 3, 6, 6

Group B: 3, 7, 1, 2, 7, 4, 3, 5, 9, 6, 3, 9, 6, 8, 3, 4, 5, 1, 8, 5, 3, 4, 8, 3, 7, 6, 6, 2, 2, 4

Determine the modal number of sneezes for each group. If Group A represented the company would you agree that they ought to change their name or have they got it right? Should one consider factors other than the mode when making such a decision?

The mean

There are three principal versions of the mean: the *Arithmetic* (pronounced with the emphasis on the e), the *Geometric* and the *Harmonic*. Their common denominator is that they can all be expressed fairly easily in mathematical terms and, hence, lend themselves to further development.

Of the three the Arithmetic is by far the most significant and widely used. I shall deal briefly with the Geometric mean and say even less about the Harmonic.

The arithmetic mean

It is at this stage that it becomes necessary to introduce some algebra and to start doing some calculations. I shall try to make these as easy as possible.

The arithmetic mean is the representative value most usually associated with the word 'average' and is so widely used that the adjective Arithmetic is usually dropped, a practice I shall follow from now on.

It is calculated by adding up all the values obtained as results of a survey and then dividing by however many there are. This is best illustrated by an example from the world of chance.

In a dice-throwing game using two dice, the following scores were obtained:

4, 1, 3, 5, 7, 6, 2, 8, 6, 7

To get the mean we first of all add them all up:

4 + 1 + 3 + 5 + 7 + 6 + 2 + 8 + 6 + 7 = 49

and divide by 10 since there were 10 results:

49/10 = 4.9

This final number is the mean. Notice that it is impossible to throw a score of 4.9. It is often the case that not merely is the mean not one of the values recorded but it is, indeed, an impossible number when referred back to the original survey – similar to the famous 2.4 children in the average family.

Exercise 4.2

1. In an infants' class the teacher asked his children how many brothers and sisters they had. These are the replies. 0, 1, 2, 1, 3, 0, 1, 2, 1, 2, 0, 3, 5, 1, 4, 1, 2, 2, 3, 1.

   Find the mean number of brothers and sisters in the class.

2. John Clarke was bored in Maths class one day and decided to measure the lengths of his fingers. He found them to be (in cm):
   6.8, 7.9, 7.2, 5.5, 6.1, 6.3, 7.2, 8.1, 7.6, 5.8.

   His teacher unfortunately caught him in the middle of this activity and, since it was a Statistics lesson, asked him to calculate the mean length of his fingers, at the same time making a rather unkind comment about it being the same as the distance between his ears. What answer should John get?

For the sake of abbreviation and ease of use, the mean is often represented by the letter m or by $\bar{x}$ (x bar). In more advanced work the Greek letter μ (pronounced 'mu') is also found. This is used to show a distinction between the mean m got from a SAMPLE of a total population and the ACTUAL mean μ of the total population. Since for our work we do not need to make this distinction, I shall restrict myself to using m or $\bar{x}$. However, an appreciation of the difference at this stage will make life a lot easier for you later on.

Example: in the ten-yearly census of Normalia it always takes rather a long time to process the results. It is standard practice, therefore, before publishing the full results, to prepare a prediction of the final outcome by a process of sampling. To this end, out of the 965 467 census forms returned, 1 000 are selected at random and processed first. Since the census forms are issued to households rather than to individuals, the 1 000 census forms are found to contain information on 3 995 people. As a result of this sample of 1 000 homes the census board publish the fact that the sample mean $m = 3\,995/1\,000 = 3.995$ people per household. When, two years later, they have finally processed all the census information, it turns out that the total population at the time of the census was 3 805 871 people. Dividing this by the number of households, 965 467, gives the actual mean μ as 3.942.

There is not a great difference between these two values and in most cases it is reasonable to use the mean m of a sample rather than to have to find the mean μ of the parent population. Indeed in most cases it is impractical to find μ and so m is very widely used. As can be seen from the census example, however, it is very rare for these two values to be exactly the same. It is therefore necessary to make a clear distinction between them.

With a calculator it is very easy to work out m. Unfortunately it can often be rather tedious to enter all the numbers. This can happen when there are a lot of them or if they are all large numbers. In the former case there is little that can be done other than perhaps grouping the data and handling it in a way which I shall explain later. With the latter problem there is frequently a short cut available called the method of the assumed or false mean.

This is used when the values are all in roughly the same range and is a method for shortening them. The idea is to make a guess at the value of the mean, call this the assumed or false mean, and then to find how far each value is from this. The mean of these differences when added to the false value will tell you the actual mean.

Example: Sergeant Marks is involved with the processing of applicants for the Normalia police force. He notices that his latest batch of twenty seems rather smaller than usual and to check on this he measures them all and finds out the mean height. His results are as follows and his calculations are shown in Table 4.1.

Heights(cm) 177,179,176,173,172,174,177,174,172,181,175,176, 174,173,169,176,182,175,181,174

Assumed mean = 176

(*Note*: The assumed mean is just a value that looks roughly in the middle. He could just as easily have used 170 or 177 or any other number.)

Table 4.1

Height (cm)	Height – assumed mean (D)	
177	1	
179	3	
176	0	
173		−3
172		−4
174		−2
177	1	
174		−2
172		−4
181	5	
175		−1
176	0	
174		−2
173		−3
169		−7
176	0	
182	6	
175		−1
181	5	
174		−2
Total	21	−31 = −10

Mean difference from assumed mean = −10/20 = −0.5.
Actual mean = 176 − 0.5 = 175.5.

Exercise 4.3

1. The air pressure over Normalia was measured on twelve successive days and found to be (in millibars)
   994, 996, 998, 1010, 1008, 1010, 1003, 1005, 1004, 1001, 998, 997
   By choosing an appropriate false mean, calculate the mean pressure over the twelve-day period.

2. Ten boys on holiday weighed themselves on a seaside weighing machine. Their weights were
   9st 12lb, 9st 10lb, 10st 4lb, 10st 9lb, 9st 6lb, 10st 3lb, 10st 2lb, 9st 7lb, 9st 13lb, 10st 10lb.
   Remembering that there are 14lb in 1 stone and using 10 stone as an assumed mean, calculate the mean weight of the boys.

Frequency tables

Because there were twenty values in our example above, the table became rather long. It could more easily have been condensed into a grouped frequency table. This also shortens the arithmetic a little. It relies on the fact that multiplication is just a short way of adding together the same number several times:

$$4 + 4 + 4 + 4 + 4 = 5 \times 4 = 20.$$

So the calculation could have been presented as in Table 4.2, giving:

Mean = 3 510/20 = 175.5

Or as in Table 4.3, using 174 as the false mean this time.

Mean Distance (D) = 30/20 = 1.5

Actual mean = 174 + 1.5 = 175.5

It is at this point that we have come to the first major source of mistakes when processing statistics. Consider the information in Table 4.4 concerning the ages of forty people in the range 20–29.

Table 4.2

Height (cm)	Frequency	Ht × Freq.
169	1	169
172	2	344
173	2	346
174	4	696
175	2	350
176	3	528
177	2	354
179	1	179
181	2	362
182	1	182
Total	20	3 510

Table 4.3

Height (cm)	Ht − ass. m (D)	Frequency	D × Freq.
169	−5	1	−5
172	−2	2	−4
173	−1	2	−2
174	0	4	0
175	1	2	2
176	2	3	6
177	3	2	6
179	5	1	5
181	7	2	14
182	8	1	8
Total		20	30

Table 4.4

Age	Frequency
20	4
21	5
22	1
23	3
24	6
25	2
26	7
27	4
28	5
29	3
Total	40

When asked to calculate the mean age of this group of people a common mistake is to add up

$$20 + 21 + 22 + 23 + 24 + 25 + 26 + 27 + 28 + 29 = 245$$

and then to divide this by 40 giving an answer of 6.125. *This is wrong!* It does, however, show one way to check your answers. Remembering the mean's definition as a representative value, an answer of 6.125 years looks rather out of place for a group of people in their twenties.

Always check that your average looks like a reasonable figure to represent the information.

Another source of error associated with this table is to add up the numbers 20 to 29 getting 245 and then to divide by 10 because there are 10 numbers. This gives the more reasonable answer of 24.5, but one that is equally incorrect.

Exercise 4.4

1. Work out the mean age for the group of people in Table 4.4, remembering that in the survey four people were aged 20 and so the number 20 must be counted four times and so on.

2/3/4. Calculate the mean value for each of the sets of data in Exercise 4.1. Use the Frequency Tables which you constructed as an aid.

One problem connected with using the arithmetic mean as a representative value is that a few very large or very small values may seriously distort the picture. Indeed Gus is not above making use of this fact . . .

Consider the case of Consolidated Holograms Ltd, a small company with ten employees including Mr Rohl the owner. The company has been doing very well for itself and the employees feel that it is about time that their pay packets reflected this. They go along to Mr Rohl and put this to him but he produces the figures in Table 4.5 and calculates:

Total monthly pay   = N$ 22 600

Mean monthly pay   = N$ 22 600/10 = N$ 2 260

Table 4.5

Name	Monthly earnings (N$)
Mr Rohl	15 000
Ms Carr	1 200
Jenny	800
James	800
Bryan	800
Karen	800
Sarah	800
Michael	800
Stuart	800
Carol	800
Total	22 600

Since the national average is only N$1 150 it is obvious that the employees of Consolidated Holograms Ltd are rather well off. Or so thinks Mr Rohl who may not be entirely unbiased in this use of statistics. His employees feel that the modal value of N$800 is a far better representative value than the arithmetic mean and they do seem to have a point.

To get around this potential for misrepresentation there are two other types of average used. The simpler, the *median*, shall be dealt with later. I am now going to explain the *geometric mean* (G.M.). The mathematics involved gets rather complicated so you may skip this section without too much harm since the G.M. is not very widely employed.

The geometric mean

Like the arithmetic mean, the geometric mean can, relatively easily, be defined mathematically and hence developed further. Unlike the arithmetic mean, it is rather hard to work out and is, hence, not very popular.

Instead of adding the values together, we multiply them and then, to get back to a representative value, we take the nth root of this answer, where n is the number of values. The calculation for the above table (Table 4.5) looks like this:

$$15\,000 \times 1\,200 \times 800 \times 800 \times 800 \times 800 \times 800 \times 800 \times 800 \times 800$$
$$= 3\,019\,898\,880\,000\,000\,000\,000\,000\,000\,000$$

Taking the tenth root of this gives a mean monthly pay for the company of N\$1 117 to the nearest dollar. (*Note*: the tenth root of a number means that number which, when multiplied by itself 10 times, gives you the original number. For example, 2 is the tenth root of 1 024 because $2 \times 2 \times 2 \times 2 \times 2 \times 2 \times 2 \times 2 \times 2 \times 2 = 1\,024$)

Obviously this gives a much better representative value than the arithmetic mean and it looks as though the employees who, by this measure, earn less than the national average of N\$1 150 have a reasonable case to make.

You may at this point wonder how the national average was calculated . . .

The issue soon becomes political and you are back to the original meaning of the word Statistics as figures about the State.

While the geometric mean irons out the effects of extreme values, its difficulty of calculation makes it rare in its appearance. I offer these key presses as a method for working out the geometric mean on a scientific calculator. There are other methods, depending on the keys available to you, but one of these should work on most makes.

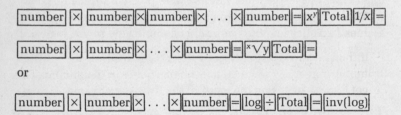

or

Here number stands for the various values and Total stands for however many there were of them, i.e. ten in my example. There is a BASIC program in the Appendix which will calculate the geometric mean for you.

Exercise 4.5

1. Calculate the geometric mean of these sets of numbers, giving your answer correct to one decimal place.

(a) 1, 2, 9.
(b) 6, 7, 8, 2, 3, 25.
(c) 41, 93, 52, 46, 47.

The median

The other average which is unaffected by extreme values is called the *median*. It is the most natural measure of the centre of a distribution and is very easily found.

Quite simply, you take all the results of your survey, arrange them in order of size and then choose the middle one. This value is the median of the distribution.

Example: a group of eleven people were found to have the following shoe sizes:

Original data 3, 5, 4, 6, 7, 8, 6, 4, 2, 4, 5

Sorted data   2, 3, 4, 4, 4, **5**, 5, 6, 6, 7, 8

Median value = 5

There are five values smaller than or equal to the median and five values larger than or equal to it. It is therefore the middle value.

If there is an even number of results from a survey then a slight complication arises since there is no one value in the middle. This is got around by taking the mean of the two central values.

Example: in an examination on the Ancient History of Normalia, a class of ten students scored the following marks (out of 100)

34 23 45 96 18 63 44 58 12 39

To find the median we first sort the data.

12 18 23 34 **39 44** 45 58 63 96

Since there are ten values, the median is the mean of the fifth and sixth values, i.e. 39 and 44.

$$\text{Median} = \frac{39 + 44}{2} = 41.5$$

There is, as with the arithmetic mean, a frequent source of error when working out the median of a distribution which is recorded as a frequency table. Table 4.6 takes as an example the same figures which I used in the section on the mean (Table 4.4).

Table 4.6. *Information concerning the ages of forty people in the range 20–29*

Age	Frequency	Cum. Freq.
20	4	4
21	5	9
22	1	10
23	3	13
24	6	19
25	2	21
26	7	28
27	4	32
28	5	37
29	3	40
Total	40	

Since there were forty values in the survey, the median value is calculated by taking the mean of the twentieth and twenty-first values. To find these you could write out the whole list in order, but this can be avoided by making use of the idea of cumulative frequency met in Chapter 2 and as recorded in the third column of the table.

If we were to write out the forty values in order, then by looking at the cumulative frequency table we could see that the nineteenth value would be 24 while the twentieth and twenty-first values would both be 25. The median age for this group of forty people is therefore the mean of 25 and 25, i.e. 25 years.

The mistake which is commonly made here is, as with the mean, to ignore the frequencies and to work out the median of the ages.

20 21 22 23 24 25 26 27 28 29,

giving an answer of 24.5 years. Be very careful not to fall into this trap.

While the median is a very natural measure for the centre of a distribution, and while it is comparatively easy to work out, it is not, unfortunately, very easily defined mathematically and so does not, unlike the mean, lend itself to much further development.

Exercise 4.6

1/2. Calculate the median values for Questions 1 and 2 of Exercise 4.3.

3/4/5. Find the median values for the data from Exercise 4.1.

6. You should now have worked out three averages for the figures in Exercise 4.1, Question 3. Taking these into consideration, and assuming that your name is Gus, an employee of the rival company, which average would you choose to represent the figures? Which average do you think is actually the most representative?

Averages of grouped data

The mode

As already mentioned, if information is presented in the form of grouped data it is customary to talk about the *modal class* as the interval which has the highest frequency. Nothing more needs to be added here.

The mean

The mean of grouped data is calculated in a similar fashion to that already described for a frequency table. Each class interval is represented by its mid-point. This is easily calculated by adding together the two boundary values and halving the answer.

Thus for a class interval 140–149cm, say, when the measurements are correct to the nearest cm, the representative value would be

$$\frac{139.5 + 149.5}{2} = 144.5\text{cm}.$$

Note that, as previously discussed, 140 actually represents a lower boundary of 139.5cm and 149 an upper boundary of 149.5cm.

The calculation of the mean of grouped data can be seen in the following example.

In an experiment a group of 100 students were asked to close their eyes and to put their hand in the air when they thought that one minute had passed. The length of time passed for each student was recorded.

Time (secs)	20–29	30–39	40–49	50–59	60–69	70–79
Frequency	3	13	29	26	20	9

Using this information, laid out as in Table 4.7, the calculation of the mean amount of time that the students had their eyes shut is

Table 4.7

Time	Mid-point M	Freq.	Freq. × M
20–29	24.5	3	73.5
30–39	34.5	13	448.5
40–49	44.5	29	1290.5
50–59	54.5	26	1417
60–69	64.5	20	1290
70–79	74.5	9	670.5
Total		100	5190

Mean $= 5190/100 = 51.9$ secs.

The median

The simplest way to find the median value of grouped data is to draw a cumulative frequency curve and then use it to read off the middle value. So, using the example above and trying to find the median amount of time, we first of all draw the graph in Figure 4.1.

If the graph were drawn extremely accurately, then the horizontal line would start at 50.5 since there are 100 entries and 50.5 is half way between the fiftieth and the fifty-first.

It is possible, and less time-consuming, to calculate the median without actually drawing a cumulative frequency curve. To do this

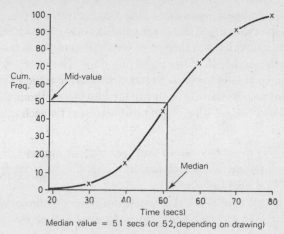

Median value = 51 secs (or 52, depending on drawing)

Figure 4.1.   Cumulative-frequency graph showing estimates of one
              minute

we decide in which interval the median lies, and then calculate how
far through the interval it comes. The calculation is done using a
cumulative frequency table (Table 4.8).

Table 4.8

Time (secs)	Freq.	Cum. Freq.
20–29	3	3
30–39	13	16
40–49	29	45
50–59	26	71
60–69	20	91
70–79	9	100
Total	100	

With 100 values, the median is the mean of the fiftieth and the
fifty-first or it can be looked on for our purposes as the '50.5th' value.
From the cumulative frequency table, there are forty-five people who
estimated 49.5 seconds or less and seventy-one who estimated 59.5
or less. This implies that the median value is in the range 49.5–59.5
seconds.

Since what we are looking for is the 50.5th entry, we deduct from

50.5 the cumulative frequency of the previous interval (i.e. the 40–49 second interval) to find that the median is 50.5–45 = 5.5 entries into the next interval. Since there are twenty-six entries in this interval the median is 5.5/26 of the way through it. The range of values is 10 seconds so the median value probably lies $10 \times 5.5/26 = 2.1$ seconds from the start. This gives a median value of $49.5 + 2.1 = 51.6$ seconds.

Exercise 4.7

1. Over a two-month period the rainfall on Normalia was as shown in Table 4.9.

Table 4.9

Depth of rain in 24 hours (nearest mm)	Number of days
0–	8
2.5–	14
5.5–	12
8.5–	9
11.5–	7
14.5–	4
17.5–	3
20.5–	0
24.5–	2
27.5–30.5	2
Total	61

1. Find (a) the modal (b) the median (c) the mean depth of rain per day over the period.

2. Sixty people were weighed prior to taking part in a programme at the Normalia Cold Cure Centre. Their weights were as follows (correct to nearest kg).
   60, 52, 54, 64, 60, 60, 49, 49, 55, 51, 58, 50, 53, 52, 61, 47, 65, 49, 57, 62, 60, 61, 53, 52, 61, 45, 57, 52, 53, 66, 50, 68, 58, 59, 67, 57, 47, 58, 54, 65, 60, 63, 53, 52, 61, 53, 59, 62, 54, 51, 48, 55, 62, 61, 55, 55, 50, 48, 44, 60
   Convert these values into a grouped frequency table taking classes 44–46, 47–49, 50–52, etc.

Use this table to obtain an estimate of the mean weight and of the median weight.

Now calculate the actual values using the raw data and see how they compare.

3. The Better Bitter Company of Normalia was contemplating launching a new high-strength lager. To test the likely success of such a product they commissioned a survey of the drinking habits of the Great Normalia public. One important question in the survey was the age of the respondent. The information was classed as in Table 4.10.

Table 4.10

Age	Frequency
18–21	243
22–26	487
27–35	184
35–50	72
50+	14
Total	1 000

Calculate an estimate of the mean and median for this sample (be careful about the class boundaries).

5 Measures of Spread

The people of Normalia are very keen on fishing. Their needs are met by Codling Fishing Supplies Ltd who manufacture, among other things, fishing line. They supply this in various strengths ranging from 1 lb to 50 lb breaking strain.

It is very important for the company's reputation that it provides a good quality, standard product. To make sure of this, the quality-control department takes daily samples of all the lines which they make. They test these to check that they have the correct breaking strain. Their results for ten samples of the 20 lb line on two consecutive days were as follows:

> Monday: 19.4, 20.8, 19.6, 19.7, 20.0, 20.3, 20.2, 19.9, 19.8, 20.3
>
> Tuesday: 18.6, 21.3, 18.1, 22.4, 20.3, 17.8, 23.2, 18.6, 21.5, 18.2

As an exercise, work out the mean values for Monday and for Tuesday using the method of the assumed mean.

Sharon Opie, the quality-control supervisor, did the above calculation and found that while only one sample had a breaking strain of exactly 20 lb, the mean for both days was exactly 20. This was spot on target but she still felt rather unhappy about Tuesday's figures.

A person who bought line of supposedly 20 lb breaking strain, which gave way at 17.9 lb would not be too pleased; nor would the customer whose line broke at 23.1 lb. Indeed, there was altogether too much variation in Tuesday's figures.

Sharon's problem was to decide what was an acceptable amount of variation from the ideal and how to measure this variation. Fortunately she had various statistical tools to help her.

The range

Probably the simplest measure of 'spread' or *deviation* (as this variation is generally described) from the centre of a distribution is the *range*. This is simply the distance from the least to the greatest value.

So for Monday the greatest value was 20.8 and the least was 19.4. This gives a range of 20.8 − 19.4 = 1.4 lb (or, to be more accurate, 20.85 − 19.35 = 1.5 lb). There is a temptation to write the range down as 19.4 to 20.8 or 19.4 − 20.8. This is wrong. It is always the difference between the greatest value and the least value. On Tuesday the range was 23.2 − 17.8 = 5.4 lb. This was almost four times as great as on Monday and provided grounds for concern.

If Sharon decided to use this as her measure of spread then she would have to decide on some acceptable range of values above which she would report the matter to the Production Manager. If she chose 1.5 lb, then Monday's figures would be passed but Tuesday's would call for some correction to the production process.

While the range provides an easy method for calculating spread, it does not give a true or complete impression of the way in which the data is dispersed.

Consider data from the census returns for Normalia. Table 5.1 shows the distribution of ages to be found in three streets in Medianville. In each case the mean age is about 43 years and the range for each may be taken as 100 years. However, it is fairly obvious that there is a great difference in the spread away from the means. In

Table 5.1

Age range	Frequencies		
	Fisher St	Kendall Rd	Gosset Ave
0–19	17	24	3
20–39	13	13	23
40–59	16	4	35
60–79	15	9	2
80–99	4	15	2
Total	65	65	65
Mean age	42.6	43.2	42.9

general the range is not a very good representative value for measuring spread since it conveys no impression of how the data is spread out in between the least and greatest values. Freak values will also distort the picture badly.

Deviation

A better measure can be obtained by finding the deviation of each statistic from the mean (i.e. how far it is from the mean) and finding the average of this.

If we take Tuesday's figures (see p.74), arranged as in Table 5.2, the calculation looks like this:

Mean 200/10 = 20

Mean dev. 17.4/10 = 1.74

Table 5.2

	Breaking strain (lb)	Deviation from mean		Absolute deviation from mean
	18.6	−1.4		1.4
	21.3		1.3	1.3
	18.1	−1.9		1.9
	22.4		2.4	2.4
	20.3		0.3	0.3
	17.8	−2.2		2.2
	23.2		3.2	3.2
	18.6	−1.4		1.4
	21.5		1.5	1.5
	18.2	−1.8		1.8
Total	200	0		17.4

This measure of spread is called the *mean absolute deviation*. It gets this rather long title from its three aspects.

Mean   = average

Absolute = ignore whether the value is positive or negative

Deviation = distance from the centre

Actually, finding the absolute value of a number merely involves changing it to a positive number if it is negative. So the absolute value of −6 is 6 and the absolute value of 12 is 12 since it is already positive. We have to do this before calculating the mean deviation. If we fail to, then the sum of the deviations always turns out to be 0, as in the example, which does not tell us a lot!

Exercise 5.1

1. Find the mean deviation from the mean for Monday's figures.

2. Ten people were asked to guess a number between 1 and 20. Their guesses were:

   2, 15, 7, 3, 18, 12, 10, 14, 15, 9

   If the number being thought of was 11, find the mean deviation from this number.

3. Ten men whose shirt size was 15 had their necks measured. The results (to the nearest tenth of an inch) were

   14.9, 14.7, 15.3, 15.1, 15.2, 15.0, 14.6, 15.4, 15.3, 15.2

   Find the mean neck size and the mean deviation from this size.

Standard deviation

Unfortunately it is, as with the mode and median, difficult to express absolute values in mathematical terms so that they can be further developed. To get around this it is more usual to do a rather complicated calculation to derive the *standard deviation* of a frequency distribution.

To calculate this the steps are as follows:

(1) Work out the deviation of each value from the mean.

(2) Multiply each value by itself (square it).

(3) Add up the squared values.

(4) Divide the answer to (3) by the number of values (the total frequency). This gives the *mean squared deviation*. This is usually called the *variance* of the distribution (abbreviated to 'var.').

(5) Finally, to get a value back into the same units as the original deviations, take the square root of the answer to (4). This final number is called the *standard deviation* (S.D.).

Going through these steps for Tuesday's figures (p. 74) arranged as in Table 5.3 should look like this:

Table 5.3

	Breaking strain (lb)	Deviation from mean	Deviation squared
	18.6	−1.4	1.96
	21.3	1.3	1.69
	18.1	−1.9	3.61
	22.4	2.4	5.76
	20.3	0.3	0.09
	17.8	−2.2	4.84
	23.2	3.2	10.24
	18.6	−1.4	1.96
	21.5	1.5	2.25
	18.2	−1.8	3.24
Total	200	0	35.64

Mean $= 200/10 = 20$

Variance $= 35.64/10 = 3.564$

S.D. $= \sqrt{3.564} = 1.89$

Before working out columns 2 and 3 it is necessary to use column 1 to calculate the mean and then to use this to calculate the deviations. Notice that, when we retain the '+' or '−' signs in the deviation, the total deviation from the mean is always going to be 0. This is because of the way we define the mean as the centre of a distribution.

I shall use the letter '*s*' to stand for the S.D. If you remember, we use *m* for the mean of a sample and μ for the mean of the parent population. It is usual to extend this practice to standard deviations. So *s* stands for the S.D. of a sample, while σ (the Greek letter s, pronounced sigma) is used for the standard deviation of the total population.

A word of warning is necessary here about the use of calculators with statistical functions when calculating the standard deviation.

The definition which I have given for the standard deviation is not necessarily the one used on your calculator. In my method at step (4) you divide by the total frequency n to get the variance. In some cases you get a more useful value by dividing instead by $n - 1$ (1 less than the total frequency). It is well beyond the scope of this book to explain this (see M. J. Moroney, *Facts From Figures*, Pelican, Harmondsworth, 1982, pp. 225ff.).

There appears to be no agreed standard on calculators with some dividing by n, some dividing by $n - 1$ and some offering both options. For our purposes we are interested only in σ_n (or rather s_n). If your calculator does not offer this as a function, then you must accept some inaccuracy (unimportant if n is large) or else multiply your answer by $\sqrt{(n-1)/n}$.

There is an alternative method for calculating the standard deviation. In many instances it is easier and it is the method I use in the program at the back of the book. It is possible to show that the S.D. can be worked out as set out here:

(1) Square each value.

(2) Add up all of these squares.

(3) Divide this result by n, the number of values.

(4) Subtract the square of the mean value to get the variance.

(5) Find the square root of (4).

The working for Tuesday's figures arranged as in Table 5.4 using this method is:

Table 5.4

	Breaking strain (lb)	Squares of values
	18.6	345.96
	21.3	453.69
	18.1	327.61
	22.4	501.76
	20.3	412.09
	17.8	316.84
	23.2	538.24
	18.6	345.96
	21.5	462.25
	18.2	331.24
Total	200	4035.64

Mean $= 200/10 = 20$

Mean of squares $= 4035.64/10 = 403.564$

Variance $= 403.564 - 20 \times 20 = 3.564$

S.D. $= \sqrt{3.564} = 1.89$

This method of calculation can be remembered easily as

variance = mean of the squares minus the square of the mean.

The two methods of calculation are often written using a mathematical formula and this seems a good enough time to meet this kind of notation. Rather dauntingly the two methods are written as

$$\text{Var}(x) = \frac{\Sigma(x_i - \bar{x})^2}{n} = \frac{\Sigma x_i^2}{n} - (\bar{x})^2$$

Here Σ (capital sigma, the Greek capital s) means add up all of the things which come after it. x_i means each value of the variable x taken in turn while $\bar{x}$ is short for the mean of the x values.

Note that to work out $403.564 - 20 \times 20$ we do the multiplying first. This is always true. Do multiplying and dividing before adding and subtracting when they appear in the same sum.

In the next exercise use both methods and check that your answers agree.

Exercise 5.2

1. Calculate the standard deviation for Monday's breaking strain figures.

2. Find the S.D. of the figures in Exercise 5.1 Question 3.

3. In eight batches of home-made wine the alcohol content (%) was

   10.8, 11.4, 11.6, 9.3, 12.1, 8.4, 9.2, 11.8.

   Find the mean, mean deviation and standard deviation for these values.

Grouped data

When the data is presented in grouped form the calculation of the S.D. is very similar to that for the mean. Using the census figures for Fisher Street (Table 5.1) in Table 5.5 we have a calculation like this:

Table 5.5

Age	Mid-point	Freq.	Deviation from mean	Deviation squared	Freq. × (dev.)^2
0–19	10	17	−32.6	1 062.76	18 066.92
20–39	30	13	−12.6	158.76	2 063.88
40–59	50	16	7.4	54.76	876.16
60–79	70	15	27.4	750.76	11 261.4
80–99	90	4	47.4	2 246.76	8 987.04
Total		65			41 255.4

Variance = 41 255.4/65 = 634.70

S.D. = $\sqrt{634.70}$ = 25.19

Note: the mid-points of the classes in this table look a little strange. If we take the 20–39 class, then this includes everyone from those who have just had their twentieth birthday up to those who are on the verge of their fortieth birthday. The class, therefore, runs from 20 to 40 and has a mid-point of (20 + 40)/2 = 30. This applies to all of the classes.

Exercise 5.3

1. By a similar method calculate the standard deviation of the ages for Kendall Road and Gosset Avenue. (Note that Kendall Road has the largest S.D. indicating that the ages are widely spread from the mean. Gosset Avenue, which has the most compact data, has the smallest S.D.)

2. A machine produces washers, supposedly of internal diameter 10mm. Over a long period of time a record (Table 5.6) was kept of its performance.

Table 5.6

Diameter (mm)	% of samples
9.5– 9.6	1
9.7– 9.8	15
9.9–10.0	48
10.1–10.2	24
10.3–10.4	9
10.5–10.6	3
Total	100

Find the mean and standard deviation of these figures.
Find also the mean deviation from the median.

3. In Chapter 2 we met some figures relating to annual earnings in Normalia. They are reproduced here as Table 5.7.

Table 5.7

Annual earnings (Normalia $)	Frequency (nearest 1000)
0– 4975	5000
4975– 7475	15000
7475– 9975	47000
9975–12475	63000
12475–14975	121000
14975–17475	76000
17475–19975	24000
19975 +	9000
Total	360000

From this table calculate an estimate for the mean and
standard deviations of the earnings.

6 Probability

'There is a 50/50 chance that if you are female and you are reading this book, then you will live to be over 75 years old.'

'The odds against Blue Monday winning the 3.30 are 12 to 1.'

'The probability of the Deviation Party winning an election in the foreseeable future is very low unless they moderate their policies.'

We come across statements like these every day in the press and on television. The branch of Mathematics which is (supposedly) behind them is called probability theory and provides us with a very powerful and sophisticated tool for making predictions.

Increasing use is being made of this topic in all walks of life. Probability provides us with the tool to move from recording and analysing data, as we have been doing so far, and to begin to make surprisingly precise predictions about the future.

Insurance companies rely for their existence and continuing prosperity on their ability to predict fairly accurately how many of their customers are going to have their cars stolen or their pet chihuahuas die of pneumonia. Government departments across the whole spectrum of their activities try to predict future trends in population movement, traffic requirements, Public Sector Borrowing, etc. Roulette players try to predict what number is going to come up next, and betting shop habitués spend a lot of time, effort and money trying to predict the outcome of a particular race.

All the above, to a greater or lesser extent, make use of probability theory. A lot of probability theory is common sense, but there are little tricks which enable the knowledgeable to gain the edge over the ignorant in its use. Casinos and insurance companies rank among the knowledgeable and tend not to be poor organizations!

The fundamental ideas are pretty straightforward, although there is plenty of scope for confusion and error. While it is usually comparatively easy to do the mathematics, this is not always the case when dealing with the real problems. When dealing with any problem to do with probabilities, the first and most vital aspect is to get a clear statement of it and then to make sure that you understand it fully. An awful lot of time, effort and veracity is wasted by people charging into these questions without thinking first.

What is probability?

The essential ideas of probability are connected with the concept of relative frequency. *'Probability' is a prediction of the relative frequency of an event.* Thus if we tossed a coin a lot of times, we would expect it to land on heads for about one-half of the time and tails for the other.

The *probability* of getting a head is $1/2$ when we toss a fair coin. In this context if we only got heads $1/4$ of the time then we might begin to suspect that the coin or the way in which it was being tossed was unfair or biased. If we toss a fair die we expect it to show a '5' about a sixth of the time. We say that the probability of getting a '5' is $1/6$ or write $P(5) = 1/6$ for short. If, however, the die is biased we may well find that $P(5) = 1/4$ or, indeed, anything that the cheat who has made it wants to engineer. There are various ways to make a die unfair. A time-honoured method is to put a small weight behind one of the numbers. This will cause that number to land on the bottom more frequently than we might reasonably expect. The die is said to be *loaded*. In our example, loading the number '2' will cause $P(5)$ to increase because '2' is on the opposite face to '5'. There are other more subtle ways of biasing dice, ranging from shaving bits off to changing the type of plastic from which they are made. Beware of big-money games with strange dice!

Probabilities, like the relative frequencies which they predict, must always lie between 0 (the event will never happen) and 1 (the event *always* happens). They are usually written as fractions – e.g., vulgar (3/4), decimal (0.75) or percentage (75%).

While it is often obvious, as in the two examples I have used, what the probability of an event is going to be, I am going to spend a little time on the theory behind this. The time spent now will be repaid when we meet more complicated examples.

To work out the theoretical probability of a particular event occurring in a trial, experiment or survey we must first of all work out the total number of possible outcomes (T) of the experiment and the number of outcomes (E) which will provide our event. If all of the outcomes are equally likely (we expect none to come up more often or less often than any other), then the probability of the event is defined as E divided by T:

$$P(Event) = E/T.$$

To illustrate this I shall look at the instrument which led to the formal development of this theory, namely the playing card.

One experiment might consist of shuffling a pack of fifty-two cards (no jokers) and looking at the top card. What is the probability that it is an ace?

There are fifty-two cards in the pack, of which we would expect no one to come up more often than any other $(T = 52)$. There are four aces in a pack so four of these equally likely outcomes would provide our Event $(E = 4)$ and the probability of the top card being an ace is

$$P(ace) = E/T = 4/52 = 1/13.$$

If we were to conduct this experiment 130 times we would expect to get an ace 1/13 of the time and turn up ten aces. Try it! I would be very surprised if the result was exactly ten but it should be fairly close, assuming that you shuffle the pack properly each time to ensure that all the outcomes remain equally likely.

Probabilities are rarely exact indicators of future events but, used properly, and with a healthy respect for their limitations, they provide a powerful tool with which to be precise about the imprecise.

Exercise 6.1

1. What is the probability of getting a vowel when you choose a letter at random from:
   (a) the alphabet,
   (b) the word 'Mississippi',
   (c) the word 'rhythm'?

2. What is the probability of selecting at random from a pack of playing cards (without jokers):
   (a) a face card,
   (b) a red '7',
   (c) a card lower than a '5' (aces high)?

3. If you toss a fair die, what is the probability that you will get:
   • (a) an even number,
   (b) a multiple of 3,
   •(c) a factor of 12,
   (d) a factor of 60?

4. In a telepathy experiment a pack of cards is used in which each card is marked with one of four symbols. There are
   • five marked with a circle, twelve with a cross, twenty with a square and thirteen with three wavy lines.
   (a) If a card is drawn at random, what is the probability that it shows a square?
   (b) If the card drawn is, in fact, a circle and it is not replaced, what is the probability that, if another card is chosen, it is not a circle?

Combined events

Sometimes the set of all possible outcomes can be confusing. There is a simple game played in quiet corners of Ireland called (from the historical times when they existed) Pitch Ha'penny. Two coins are tossed into the air simultaneously and people bet on the way they land. On the surface there appear to be three equally likely outcomes:

(a) two heads;

(b) two tails;

(c) one head and one tail.

These outcomes give probabilities of 1/3, 1/3 and 1/3 respectively.
If you ever get asked to play this game, then do not work on these
odds! If you are not sure why, then try playing the game a few times
and work out the *actual* relative frequencies or else run the program
in the appendix. The results will not, in the long run, tie in with
these predictions.

In actual fact there are four possible outcomes to this experiment,
not three. If we call the coins A and B, we can draw up a table of
outcomes like Table 6.1 to show what I mean.

Table 6.1

	Coin A	Coin B
Outcome 1	H	H
Outcome 2	H	T
Outcome 3	T	H
Outcome 4	T	T

The probabilities are therefore:

$$P(2 \text{ heads}) = 1/4$$

$$P(1 \text{ head}, 1 \text{ tail}) = 2/4 = 1/2$$

$$P(2 \text{ tails}) = 1/4.$$

As backgammon or Monopoly players know, another common
occurrence of this type of situation is when two dice are thrown and
their dots added together. There are eleven possible events. The
totals may be any one of 2, 3, 4, 5, 6, 7, 8, 9, 10, 11, 12; but P(2) is not
1/11.

If you are not convinced of this, try throwing two dice seventy-
two times and record the relative frequencies of the eleven events.
Alternatively run the simulation program in the appendix.

Again, here, the two dice act separately. Imagine that one die is
red and the other is blue. The outcomes can be recorded in a table
like Table 6.2.

Table 6.2

R	B	R	B	R	B	R	B	R	B	R	B
1	6	2	6	3	6	4	6	.	.	.	.
1	5	2	5	3	5	.	.	.	.	.	.
1	4	2	4	.	.	.	.	.	.	.	.
1	3	.	.	.	.	.	.	.	.	.	.
1	2	.	.	.	.	.	.	.	.	.	.
1	1	.	.	.	.	.	.	.	.	.	.

Copy and complete this table. Put rings around all the outcomes which give a total of seven and hence work out the probability of throwing a seven, remembering that $T = 36$. How does this compare with what actually happened when you threw the dice?

It is, unfortunately, pretty tedious doing these experiments enough times to check our predictions. You are always welcome to do so, but I hope that from now on you can accept on trust or intuitive understanding what I am saying.

The method which I showed you for the two dice can be used for any similar situation where there are two or more things combining to produce an event. All that is necessary is for us to list all the equally likely outcomes and then to count how many of these (E) give us the event in which we are interested.

Tossing four coins

The problem: four coins are tossed into the air. What is the probability that there are:

(a) exactly three heads showing when they land;

(b) no more than two heads showing?

The solution: there are the sixteen possibilities shown in Table 6.3.

Table 6.3

HHHH	THHH *	HTTH +	THTT +
HHHT *	HHTT +	THTH +	TTHT +
HHTH *	HTHT +	TTHH +	TTTH +
HTHH *	THHT +	HTTT +	TTTT +

(a) Of these sixteen possible outcomes four (marked *) give exactly three heads. So P(3 heads) = 4/16 = 1/4 = 0.25 = 25%. (The probability may be written in any of these three forms – a vulgar fraction, a decimal fraction or a percentage.)

(b) There are eleven ways of two or fewer heads occurring (marked +). So P(2 or fewer heads) = 11/16 = 0.6875 = 68.75%.

I would like to stress again here that this means that, if we toss four coins a large number of times, the relative frequency of two or fewer heads is going to be very close to 11/16. In other words, out of 16000 tosses roughly 11000 would be expected to produce two or fewer heads. 11000 is unlikely to be the exact figure but it should be close and there are powerful techniques available (see Moroney) to predict how close you would expect the answer to be to the predicted 11000. Again this answer would be given in the language and terms of probability theory.

Exercise 6.2

1. If three coins are tossed simultaneously, what is the probability of getting exactly two heads?

2. A game of Snap is played between two friends. They each have a full pack of cards. As usual they take it in turn to flip over a card. What is the probability that the first two cards shown are
    (a) both hearts,
    (b) both the same suit,
    (c) different suits?

3. In a game of Dungeons and Dragons special dice are used. One variety has eight faces, numbered from 1 to 8. If two such dice are thrown together, what is the probability that the total score showing is going to be:
    (a) eleven;
    (b) odd;
    (c) greater than twelve;
    (d) less than or equal to six?

Tree diagrams

Writing out combination tables can soon become extremely time-consuming and boring. This is particularly the case if the two (or more) events being looked at can be arrived at in lots of different ways. Taking one card from each of two packs, as in the question about Snap, and looking for the same spot value would require a 13 by 13 = 169 entry table. Fortunately a device called a 'tree diagram' can make life easier for us.

This gets its name from its construction which resembles the branching of a tree. It is easier to understand if we think of it as a maze through which we want to find our way. Each time we come to a junction we choose our path by carrying out a probability experiment.

Suppose we want to find the probability, when throwing two dice, of getting double-six. Our maze will look, initially, like Figure 6.1. We imagine that at each junction we toss a die. If it shows a six we go left. Otherwise we go right. This still, though, does not tell us about the probability which we want to work out.

As a first step, imagine that we enter the maze 360 times. Because P(six) = 1/6, we expect to get to A about 1/6 of 360 = 60 times, and to B the other 300. Of the 60 times that we get to A, we expect 1/6 of 60 = 10 to end up at C while the remaining 50 leave us at D. If we put these frequencies on the tree diagram, and complete the process, it

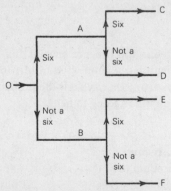

Figure 6.1.

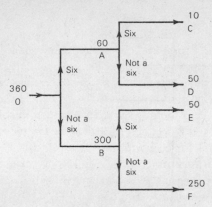

Figure 6.2.

looks like Figure 6.2. So 10 out of the 360 entries into the maze correspond to getting a six twice and hence

P(two sixes) = 10/360 = 1/36,

as we would expect from previous work.

I knew here what the final result was going to be and so I was able to choose to enter the maze 360 times in order to make the arithmetic easier and the idea clearer. If we enter the maze 100 times, it is not so easy to put actual frequencies on the diagram. To get around this, and because we want them in the end anyway, we normally put relative frequencies, or the probabilities which predict them, on the diagram instead. The final tree diagram now looks like Figure 6.3. So we can read off

P(two sixes) = 1/36

and

P(exactly one six) = 5/36 + 5/36 = 10/36 = 5/18.

We add the two results here since there are two routes which provide us with exactly one six – OAD and OBE.

We do not have to put letters on a tree diagram and they really are not as difficult as that explanation makes it seem. A more normal example might look like this:

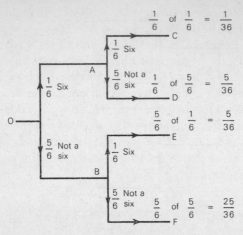

Figure 6.3.

The problem: two cards are taken from a pack without replacing them. What is the probability that at least one of them is an ace?

The solution: Referring to Figure 6.4.

P(at least 1 ace) = 12/2652 + 192/2652 + 192/2652 = 396/2652 = 0.149 (to three significant figures)

Note that 'of' is replaced here by × since this is the actual sum which we do to work out a fraction of another fraction.

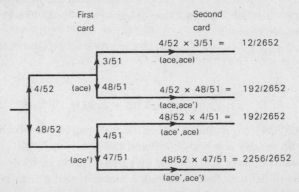

Figure 6.4.

I have here adopted two conventions. The first is the use of a dash ' to show the event *not* happening. Ace' means that an ace did not show up. Similarly, in the previous example 6' would have meant that the throw resulted in other than a 6 showing.

The other convention deals with the way in which probabilities are written. In this example I have given my answer as a decimal correct to three significant figures (s.f.). This is usually done in cases like this where the final answer works out to be a rather complicated fraction. If it is a simpler fraction like 11/16 or 4/25 then it is normal practice to leave it in that form. This is just a matter of style and in no way affects the validity of the results.

Notice that

$$P(\text{no aces}) = 2\,256/2\,652 = 0.851 \text{ (three s.f.)}$$

and that

$$P(\text{at least one ace}) + P(\text{no aces}) = 0.149 + 0.851 = 1.$$

This must be true for all tree diagrams and provides a check for accuracy. *At any stage, the totals of all the branches of a tree diagram must be one.* Using the maze analogy again, this means that we must be at one of the end points [P(reaching the end) = 1]. This can be used to simplify problems. In this case we could have said that

$$
\begin{aligned}
P(\text{at least one ace}) &= 1 - P(\text{no aces}) \\
&= 1 - 2\,256/2\,652 = 396/2\,652 \\
&= 0.149 \text{ (three s.f.)}
\end{aligned}
$$

A tree diagram can be extended to as many layers as we want. If we go back to the problem of tossing four coins we get a tree like that in Figure 6.5. And so, from all the results giving exactly three heads (marked *), we get

$$
\begin{aligned}
&P(\text{three heads}) \\
&= 1/16 + 1/16 + 1/16 + 1/16 = 4/16 = 1/4.
\end{aligned}
$$

A diagram like this is complicated to draw and can be complicated to follow. Before embarking on one make sure that the question warrants it. Very often we do not need to draw all of the branches as some of them may stop or be of no real interest with regard to the problem under consideration. Having decided to draw a tree

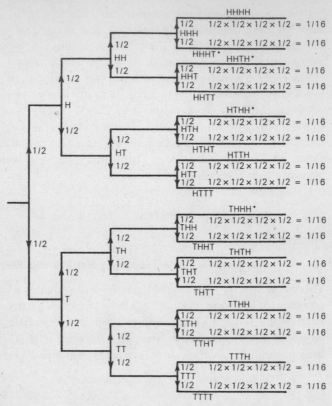

Figure 6.5.

diagram it is vital that you do so in an organized and neat fashion. An untidy tree will just lead to mistakes and wasted time in the long run. It is worth taking time over the drawing.

Exercise 6.3

1. A mother with a rare skin complaint knows that there is a probability of 1/4 that she will pass it on to any child she has. If she has a family of three, what is the probability that

   (a) none of them have the complaint,

(b) all of them have it,

(c) at least one of them has it?

2. Norman, who fancies himself as a flashy dresser, has a wardrobe of five singularly loud ties. If he wears one on a given day the probability that he will wear it on the following day is 1/9, while the other ties will each have a probability of 2/9 of appearing. On Monday he wore his orange one with the pink stripes. What is the probability that

(a) he will wear the tie on Wednesday;

(b) he will not wear it on Thursday?

3. The probability of a rainy March day in Normalia is 2/5. What is the probability that over a four-day period it rains at least once?

4. A bag contains five white balls, three red balls and two blue ones. A ball is selected at random from the bag and not replaced. This process happens three times. What is the probability that

(a) all three were red;

(b) there was one of each colour;

(c) none of them were red;

(d) there were no blue balls left in the bag?

Obtaining probabilities

So far we have dealt with what are known as *a priori* or theoretical probabilities. This just means that we worked them out from first principles and our feel for the situation. In most cases, however, this does not really work. Thus, there is no justification for the statistician employed by Acme Light Bulb (Normalia) Inc. to assert that a particular light bulb has a probability of 0.3 of lasting more than 500 hours under normal usage unless she has done some research to back this up.

There is nothing inherent in the nature of light bulbs to make this statement seem necessarily true (or false). The way in which such a probability is obtained is by testing a large number of bulbs to destruction and then calculating the relative frequency with which bulbs lasted more than 500 hours. This is the most common way in which probabilities are obtained and is called *a posteriori* probability because we work it out *after* experimenting.

The important thing to remember is that no matter how the probabilities are initially obtained, the subsequent work on combining them is the same. So that if two bulbs were taken at random from the batch talked about above, and assuming the given *a posteriori* probability to hold,

P(both bulbs last more than 500 hours) =
$0.3 \times 0.3 = 0.09$.

Exercise 6.4

1. Joan Summers breeds Burmese cats. She has a Brown stud and one of her queens is a Blue. After they had produced six litters between them she noted that of twenty-five kittens born, fifteen were brown, six were chocolate and four were blue. The next litter had three kittens. What is the probability that
   (a) the first born was
      (i) brown, or
      (ii) chocolate, or
      (iii) blue;
   (b) there was not a blue in the litter;
   (c) all three kittens were brown?

2. A party of marine biologists were studying the habits of two types of sea snail. They searched 500 rock pools looking for the two types of snail and found

Pools with neither	54
Pools with A only	156
Pools with B only	196
Pools with both	94

   (a) How many pools contained A?
   (b) What is the probability that a pool chosen at random contained type A?

(c) What is the probability that a pool chosen at random contained type B?

(d) By combining the results of (b) and (c), and assuming them independent, what would you expect the probability to be of finding a pool, chosen at random, containing both types?

(e) What was the actual relative frequency for this?

(f) Does this support the hypothesis that the two types of snail do not get on with each other?

Dependence and independence

Question 2 of Exercise 6.4 illustrates a grave potential danger when combining probabilities. In all of the examples which we have met up to this one, it has been reasonable to say that

$$P(A \text{ and } B \text{ happening}) = P(A) \times P(B).$$

While this is true in many situations, it is not universally so. We can only use this result if event A has no effect on event B and vice versa. If this is the case, then A and B are said to be *independent*.

The notion of independence is very important when probability theory is applied to research. To see this in action let us look at the investigations of the Normalia Hair Research Institute into the efficiency of Mohair, a product which claims to restore hair to bald men. The Institute's clinical trials produced the results shown in Table 6.4.

Table 6.4

Men reporting			
Treatment	Restoration	No restoration	Total
Mohair	80	420	500
Placebo	40	460	500
Total	120	880	1 000

This table is first replaced with one showing relative frequencies rather than actual frequencies. They are written as decimals in Table 6.5 for convenience of calculation and comparison.

Table 6.5

Men reporting			
Treatment	Restoration	No restoration	Total
Mohair	0.080	0.420	0.50
Placebo	0.040	0.460	0.50
Total	0.12	0.88	1.00

Now we consider the case of a man chosen at random from the sample of 1000. From the right-hand column

P(man was treated with Mohair) = 0.50

From the bottom row

P(man showed hair restoration) = 0.12.

If we make what is called a *null hypothesis*, that Mohair has no effect on hair restoration (the events are independent), then we are justified in saying that P(man was treated with Mohair and showed hair restoration) = P(man was treated with Mohair) × P(man showed hair restoration) =

0.50 × 0.12 = 0.06.

However the actual, observed relative frequency was 0.08. This is higher than the predicted result. It seems as though our null hypothesis has led us astray and so we should reject it and say that Mohair does, indeed, have some restorative effect (the two things are *dependent*). Unfortunately Mohair is not yet available in this country . . .

The decision to reject the null hypothesis can, and should, be made in a more informed and reasoned manner than this. Further developments of probability theory (see Moroney, Chapter 15) allow us to develop this research technique to a high level of sophistication.

The implication of experiments such as this for our work on tree diagrams is that instead of saying, as we simplistically have done up till now, that

$$P(A \text{ and } B) = P(A) \times P(B),$$

we must say

$$P(A \text{ and } B) = P(A) \times P(B \text{ given that } A \text{ has happened}).$$

This assumes that A can be viewed as happening before B. $P(B$ given that A has happened) is usually written as $P(B|A)$.

If $P(B|A) = P(B)$, that is

$$P(A \text{ and } B) = P(A) \times P(B|A) = P(A) \times P(B)$$

then we say that A and B are independent. Otherwise A and B are dependent. Going back to the Mohair example, P(man treated with Mohair and showed restoration) = P(man treated with Mohair) $\times$ P(man showed hair restoration given that he used Mohair). That is, $0.08 = 0.50 \times$ P(man showed hair restoration given that he used Mohair). This tells us that P(hair restoration given that Mohair is used) $= 0.08/0.50 = 0.16$.

What Gus makes of this kind of manipulation and subsequent wording is the stuff of which advertising campaigns are made.

Exercise 6.5

1. A card is drawn from a pack of playing cards and not replaced. A second card is then drawn. If A is the event that the first card is a seven and B is the event that the second card is a seven, what is
   (a) $P(B|A)$,
   (b) $P(B'|A)$,
   (c) $P(B|A')$,
   (d) $P(B'|A')$
   (e) $P(AB)$ (this is short for the probability of A and B happening),
   (f) $P(AB')$
   (g) P(getting exactly one seven in the two cards)?

2. In a survey of 1 000 students at the University of Normalia, 120 professed themselves to be anarchists, 400 admitted

having overdrafts with their bank, while of the anarchists 70 had overdrafts. If a student was chosen at random from the 1 000 surveyed, what would be the probability that she was

(a) an anarchist;

(b) overdrawn;

(c) an anarchist and overdrawn?

Are these two events dependent or independent?

7 Weighted Averages and Index Numbers

'All-Share Index down 4.39 at 582.21.'

'Sterling Index 73.0 (1975 = 100).'

'Retail Price Index 358.8 (November) up 4.9 per cent on year.'

'Hang Seng Index up 11.11 at 1184.42.'

Have you ever wondered to what these mystic numbers refer? The papers and TV use them as if they are meant in some way to reflect the state of a country and its economy. It seems that, in general, upward movement is a good thing (except for the Retail Price Index). Unfortunately, although widely quoted, there is rarely any explanation offered as to their exact meaning and formulation. While I do not propose to explain how individual indices are obtained, I hope to offer a sufficient overview to enable you to find out easily for yourself.

Indices

The idea of an index in this context is, in principle, very simple although the actual calculation may become complicated and, in the case of the Retail Price Index, charged with political overtones.

In essence an index of this type is just a fancy percentage. It usually compares the way things stand at one point in time with the way in which they stand or stood at another. So, if processed peas cost

N$0.80 in January 1986 and N$0.90 in January 1987, the *index* for processed peas with January 1986 as *base* is

$$\frac{0.90}{0.80} \times 100 = 112.5.$$

And that is really all there is to it.

The Sterling Index (base 1975) quoted above compares the value of the pound now against its value in 1975. Suppose that in 1975 it cost £x to buy 100 units of a mixture of the major foreign currencies. We could now, on average, buy only 73.0 units for the same money. This is a simplistic account of the actual calculation since the index takes into account the relative importance, in terms of trade, of the currencies as well.

The formula for calculating an index is

$$\text{Index} = \frac{\text{value}}{\text{base value}} \times 100.$$

The value under consideration may be anything of a numerical form that changes with time. It may be the number of cars produced in Normalia or the population of cocker spaniels in Liechtenstein. The common theme is that the index provides us with a simple, uniform method of comparing changes.

Index numbers are often called *relatives* because they describe the way in which two commodities relate to each other. So an index of prices provides price relatives while an index of industrial output would provide industrial output relatives.

Exercise 7.1

1. Here are several pairs of values which are connected. For each pair calculate, to one decimal place (d.p.), an index number
   (a) taking the first value as base,
   (b) taking the second value as base.

	(i)	(ii)	(iii)	(iv)	(v)
First value	N$120	3.2km	150kg	£470	90min.
Second value	N$150	2.5km	180kg	£429	250min.

It is fairly natural, if an index has been running for a very long time, to want occasionally to change the base year or date. A Sterling Index with 1890 as base year would not be of much relevance today. The process involved is, again, fairly straightforward.

As part of its annual report Codling Fishing Supplies Ltd produces an index of its performance over a twelve-month period (Table 7.1). As you can see, the company's value dipped mid-year but rallied well at the end to finish twenty-three points up over the year. 'Points' is the commonly used word for the movement of indices. These indices could just as easily have been presented with December as the base so that our table would look a little like Table 7.2. Our problem is to work out what ? and ?? ought to be. The key lies in what has happened to the 123 for December.

Table 7.1. *Index of share prices for Codling Fishing Supplies Ltd (Jan = 100)*

Jan	June	Dec
100	84	123

Table 7.2

Jan	June	Dec
?	??	100

To change 123 into 100 we divide by 123 and multiply by 100. So this is what we do to the other two values as well.

$$? = \frac{100}{123} \times 100 = 81.3$$

$$?? = \frac{84}{123} \times 100 = 68.3$$

Notice that this is exactly the same calculation as the original one to find an index and so is nice and easy to remember.

Exercise 7.2

1. Given here are a set of index numbers illustrating production at the Flotsam Shipbuilding Works in Normalia (1903 as base).

1903	1904	1905	1906	1907
100	110	130	120	135

   (a) Change the base year for these figures to (i) 1905, (ii) 1907.

   (b) Change the figures into a *chain base index*. This means that for 1904 you take 1903 as base, for 1905 you take 1904 and so on.

2. Against a basket of major currencies an index is calculated showing the relative value of the Normalia $.

Year	1905	1906	1907	1908	1909	1910
Index	100	85	93	106	97	99

   Given that these figures have been calculated as a chain base index, recalculate them with

   (a) 1905 as base,
   (b) 1910 as base.

Weighted averages

In fact, while that part is easy, the calculation of something like a retail price index is much more complicated. What this is supposed to represent is an index of the change in the *average* cost of living in a country. It is the definition of 'average' which causes the problem.

Suppose that in a given year the price of potatoes has gone up by 7 per cent but that the price of asparagus has gone up by 15 per cent. Taking these two factors into account, is it reasonable to say that the average price rise for vegetables is

$$\frac{15 + 7}{2} = 11 \text{ per cent?}$$

Unless we have spent the year on a very expensive diet, it is fairly obvious that what has happened to the price of potatoes is of much more significance than the change in the price of asparagus. Some form of *weighting* is necessary to take account of the fact that people tend to spend a lot more on potatoes than on asparagus.

A simple way to balance things is to estimate how much is spent on potatoes (N$60 say) and how much is spent on asparagus (N$1 say) by the 'average' person in Normalia in a year. The '60' and the '1' then provide the *weightings* for calculating what is called a *weighted average*. This is worked out as follows:

$$\frac{60 \times 7 + 1 \times 15}{60 + 1} = 7.13 \text{ per cent.}$$

This is a much more reasonable average and reflects the relative importance of the potato price compared with that of the asparagus.

Example: as well as the Sterling Index there is a Normalia $ index. This is calculated as a weighted average of the relative changes against Normalia's five most important trading partners. The figures for a typical year are shown in Table 7.3.

Table 7.3

Currency	Change %	Weighting
Abnormalia $	+5	1.5
Yan	−9	2.0
Krumpan	+7	0.8
Plotyl	+3	1.2
Krupyk	−16	0.6

Weighted average change =

$$\frac{1.5 \times 5 - 2.0 \times 9 + 0.8 \times 7 + 1.2 \times 3 - 0.6 \times 16}{1.5 + 2.0 + 0.8 + 1.2 + 0.6} = -1.79\%$$

So Normalia $ index = 100 − 1.79 = 98.21

The rule for working out weighted averages is very similar to the rule for working out the mean of grouped data.

$$\text{Weighted average} = \frac{\text{sum of (weights} \times \text{values)}}{\text{sum of weights}}$$

In these examples I made up the weightings and shall happily do so in all future work. How weightings are derived is a matter for much manipulation and contention and one which I gladly leave to other authors.

Exercise 7.3

1. When grading a candidate, an examining board takes into account five factors to which they ascribe different weightings. These are

Area	Course work	Project	Oral	Exam paper 1	Exam paper 2
Weighting	4	2	1	2	3

The marks (per cent) for four candidates are shown in Table 7.4. Calculate, using the given weights, the average mark for each candidate.

Table 7.4

Area	John	Jane	Claude	Cecilia
Course work	47	53	69	24
Project	65	23	82	46
Oral	48	39	74	44
Paper 1	53	48	92	51
Paper 2	19	54	64	84

2. Product Magazine produces regular comparative reports on a range of products. It gets these reports by choosing a series of attributes and asking a panel of five judges to award a mark between 1 and 10 for each of them. For each attribute the marks are totalled and a final overall number is obtained by calculating a weighted average of these totals. Table 7.5 shows the weightings and the results for three brands of ice-cream. Calculate, for each brand, the final score.

Table 7.5

| | | Total marks | | |
Attribute	Weight	Brand A	Brand B	Brand C
Taste	7	25	36	19
Colour	2	41	47	12
Texture	3	28	42	21
Smell	1	40	48	15
Packaging	2	12	23	46

Weighted or composite indices

To calculate a weighted index all that is necessary is to find a weighted average at the base time and a weighted average at the current time and make an index out of these two values.

Example: the Codling Fisheries Supplies Ltd employs different grades of workers, each grouping being paid on different pay scales and receiving differing wage increases from year to year. The figures for last year and for this year are shown in Table 7.6. Calculate an index for this year's wage bill with last year as base.

Table 7.6.   *Average monthly pay (N$) for group*

Type of worker	Number employed	Pay last year	Pay this year
Management	10	2700	3000
Supervisor	15	1800	2000
Skilled	40	1800	1900
Semi-skilled	35	1500	1650
Unskilled	60	1200	1380
Total	160		

Solution:

Weighted average for last year =

$$\frac{10 \times 2700 + 15 \times 1800 + 40 \times 1800 + 35 \times 1500 + 60 \times 1200}{10 + 15 + 40 + 35 + 60}$$

$$= 1565.625$$

Weighted average for this year =

$$\frac{10 \times 3\,000 + 15 \times 2\,000 + 40 \times 1\,900 + 35 \times 1\,650 + 60 \times 1\,380}{10 + 15 + 40 + 35 + 60}$$

$$= 1\,728.4375$$

$$\text{Index} = \frac{1\,728.4375}{1\,565.625} \times 100 = 110.4.$$

Notice that in this case the number of employees in each group made the weighting for that group. To calculate an index figure like this it is not strictly necessary to divide by the total of the weights as these values cancel out in the final division but you might well find it less confusing to do it anyway.

The management of the company have worked out that the index for turnover for this year with last year as base is 113.5. This means that the amount of money coming in has increased at a faster rate than the outgoings on wages and so, unless the cost of materials and other overheads have done worse, the employees have earned a production bonus for the year.

To make my explanation easier, in this example I assumed that the number of employees remained unchanged over the year. In practice this is rarely the case. Weightings for any index need to be constantly revised due to changes in relative importance of the items. If the weightings are different for the two periods under consideration, we have to make a decision about which to use. The obvious one of using the relevant weightings at each time makes the two weighted averages less easily comparable so some compromise is usually struck. The details of this are fairly complicated and are dealt with in more detail in Yeomans.

Exercise 7.4

1. At the beginning of a school year the children of a particular class investigated the pattern of spending in their area. Each of six items was given a certain weight and an initial price relative of 100. During the school year various price rises occurred. The weights and the percentage increases in prices are shown in Table 7.7.

Table 7.7

Item	Weight	Increase in price (%)
Food	350	25
Household goods	110	21
Clothing	95	12
Housing	90	8
Fuel and light	65	18
Miscellaneous	100	13

What is the value of the 'cost of living' index at the beginning, and at the end, of the school year?

2. The records of a certain small investor include the data given in Table 7.8.

Table 7.8

		Market price in pence	
Company	Number of shares held	Nov. 1982	Nov. 1974
British Petroleum	30	520	268
Commercial Union	100	222	74
ICI	60	275	153
Marks and Spencer	60	289	122

Calculate
(a) an index for the total value of the investor's shares in November 1982 as compared with November 1974,
(b) an index for the value of the Commercial Union shares in November 1982 taking November 1974 as base.

3. In 1975 a girl spent half her weekly allowance on clothes, one-third on travelling and one-sixth on sweets and chocolate. The price relatives comparing 1976 with 1975 for these items were 105, 120 and 110 respectively. Calculate the girl's cost-of-living index for 1976 compared with 1975, assuming that her spending habits were unchanged.

4. Given that 102 is the combined index for the commodities below, find the value of *x*.

Commodity	P	Q	R	S
Index	111	107	103	98
Weight	1	3	4	x

Standardized death rates

This rather macabre section heading provides another example of the need for, and the use of, weighted averages.

For better or for worse, the government collects and publishes statistics on the number of deaths per electoral ward in Normalia. For a particular year Wilcoxon ward had 1496 deaths while Kolmogorov ward only had 636.

Norman Andrew Ive, a freelance journalist, was preparing an article on the relative healthiness of different areas of Normalia and took these figures to show that Kolmogorov was much healthier than Wilcoxon. Fortunately Norman had a friend, Caroline Lever, who always read his copy and who pointed out that these raw figures were meaningless without knowing the relative populations of the two wards. Not taken aback, Norman found that Wilcoxon had a population of 46000 while Kolmogorov had 78000 inhabitants.

To make the comparison obvious even to Caroline he decided to standardize these figures. Instead of working out the percentage death rate, however, he worked out the number of deaths per 1000 people. This is the way in which such information is usually presented. His calculations were:

$$\text{Wilcoxon} \qquad \frac{\text{Deaths}}{\text{Population in thousands}} \quad \frac{1496}{46} = 32.52$$

$$\text{Kolmogorov} \qquad \frac{636}{78} = 8.15 \text{ deaths per 1000}$$

32.52 is such a significantly worse death rate than 8.15 that Norman got quite excited and felt that he was on the verge of a major scoop. Caroline, rather more level-headed than he, pointed out that the

tremendous difference could perhaps be accounted for by the fact
that Kolmogorov was a relatively newly developed area while Wilcoxon was famed as a retirement spot. Perhaps the fact that there
were more old people in Wilcoxon explained why more people were
dying without anything more sinister going on. Norman concurred,
rather sadly, and went in quest of further statistics. He returned with
the figures in Table 7.9.

Table 7.9

	Wilcoxon		Kolmogorov	
Age	Population	Deaths	Population	Deaths
0–5	1800	23	21000	294
6–15	1200	1	14300	29
16–25	2500	8	15600	62
26–35	1500	5	12300	55
36–45	2100	7	7500	30
46–55	4300	46	4600	55
56–65	8600	206	1300	34
66+	24000	1200	1400	77
Total	46000	1496	78000	636

While the *crude death rates* for the two wards were 32.52 and 8.15
per 1000 respectively, the age distribution indicated by this table
suggested to Norman that Caroline had a point, and so he set about
processing them further. The first step was to express the deaths in
each age group as rates per 1000 as he had already done with the
totals. For Wilcoxon this came out as shown in Table 7.10.

Table 7.10.   *Wilcoxon*

Age	Population	Deaths	Deaths per 1000
0–5	1800	23	12.8
6–15	1200	1	0.8
16–25	2500	8	3.2
26–35	1500	5	3.3
36–45	2100	7	3.3
46–55	4300	46	10.7
56–65	8600	206	24.0
66+	24000	1200	50.0
Total	46000	1496	

Work out the same figures for Kolmogorov ward. As you can see, in every category the death rate was worse here than in Wilcoxon indicating that, contrary to his previous hypothesis, Norman would now have to report how unhealthy it was to live in Kolmogorov. Unfortunately the editor to whom he was thinking of selling the piece hated to see tables of figures taking up valuable column-inches and so Norman had to condense his results into one statistic.

Again Ms Lever came to his assistance and suggested that, by suitably weighting the figures, Norman could arrive at averages for the two wards which were comparable without being misleading. She proposed that he find the national distribution of population among the differing age groups and use these as weights to produce a *standardized death rate*.

This he duly did and his results for Wilcoxon are recorded in Table 7.11.

Table 7.11. *Wilcoxon*

Age	Deaths per 1 000 (R)	National population (%) (W)	W × R
0–5	12.8	10.0	128.00
6–15	0.8	15.9	12.72
16–25	3.2	11.8	37.76
26–35	3.3	12.3	40.59
36–45	3.3	13.9	45.87
46–55	10.7	10.1	108.07
56–65	24.0	14.6	350.40
66+	50.0	11.4	570.00
Total		100.0	1 293.41

So

$$\text{Average death rate} = \frac{\text{sum of weights} \times \text{values}}{\text{sum of weights}}$$
$$= 1293.41/100$$
$$= 12.93 \text{ deaths per 1 000 (two d.p.)}$$

Putting all of this together in one table produced figures like those in Table 7.12 for Kolmogorov Ward.

Table 7.12.   *Kolmogorov*

Age	Population	Deaths	Deaths per 1000 (R)	National population (%) (W)	W × R
0–5	21 000	294	14.0	10.0	140.00
6–15	14 300	29	2.0	15.9	31.80
16–25	15 600	62	4.0	11.8	47.20
26–35	12 300	55	4.5	12.3	55.35
36–45	7 500	30	4.0	13.9	55.60
46–55	4 600	55	12.0	10.1	121.20
56–65	1 300	34	26.2	14.6	382.52
66+	1 400	77	55.0	11.4	627.00
Total	78 000	636		100.0	1460.67

$$\text{Standardized death rate} = 1460.67/100$$
$$= 14.6 \text{ deaths per } 1000 \text{ (one d.p.)}$$

It was now clear that the death rate in Kolmogorov ward was significantly worse than that for Wilcoxon. Norman's next step in producing this story was to get on the phone to his great friend and adviser Gus. The rest, as they say, is history (unfortunately too often the case).

To summarize then. If we are studying a population, the *crude death rate* = Total number of deaths/population in thousands. *Age specific death rates* are the death rates (per 1000) of each chosen age group within the population. The *standardized death rate* is a weighted average rate for the population where each age specific death rate is weighted with the percentage that group represents of the national or of some standard population.

Exercise 7.5

1. Table 7.13 gives a breakdown by age of the population and deaths in one year for two towns. Calculate for each town
   (a) a crude death rate
   (b) age specific death rates
   (c) a standardized death rate.

Table 7.13

Age range	Town A Population	Deaths	Town B Population	Deaths	National Population (%)
0–5	48000	542	9300	125	6.3
6–15	57100	51	14600	11	14.7
16–25	61300	152	25100	85	11.3
26–35	84400	260	17200	45	15.8
36–45	42900	129	18300	48	17.6
46–55	38200	369	11400	110	14.3
56–65	24700	550	23600	538	11.6
66+	18600	1560	29400	1440	8.4

2. Kolmogorov ward has a reputation for a high birth rate compared with the national average which is 84.6 live births per 1000 population. A breakdown of the relevant figures for Kolmogorov ward is given in Table 7.14. By calculating a standardized birth rate, decide whether Kolmogorov deserves its reputation.

Table 7.14

Age range of mothers	Female population	Live births	National population (%)
0–5	10600	0	10.0
6–15	7500	30	15.9
16–20	4300	860	5.8
21–25	4700	1743	6.0
26–30	2700	925	6.2
31–35	2400	278	6.1
36–45	3800	143	13.9
46–55	2200	0	10.1
56–55	700	0	14.6
66+	800	0	11.4

8 Connections

*There is a very strong correlation between the stork popu-
lation of Sweden and the live birth rate in England.*

*There is a strong connection between smoking and death
by lung cancer or heart disease.*

*Having measured the period of pendulums of differing
lengths and plotted T against √l, the line of best fit con-
firms the theoretical relation $T = 2\pi\sqrt{(l/g)}$.*

These are examples of areas where we collect two sets of data and
then look for a connection between them. Such collections of data
are often referred to as *bivariate data* because there are two variables
involved in the investigation. This chapter is about ways of dealing
with the connections between data.

Scatter diagrams and lines of best fit

The Big Bang Firework Corporation wanted its bangers to
make as much noise as possible for the least amount of money. The
expensive ingredient was the gunpowder, and so John Noble, their
research officer, experimented with changing the thickness of the
cardboard tube while not changing the amount of powder. The
results of part of his research are shown in Table 8.1.

Quite obviously from looking at this data we can see that, in
general, the thicker the cardboard, the louder the bang. To make the
point more evident we can draw a graph to illustrate the data (Figure
8.1).

When drawing this graph we had to make a decision about which
variable should go across and which one should go up. This can be

Table 8.1

Thickness of cardboard (mm)	Noise (decibels)
0.5	101.1
0.7	102.0
1.0	101.8
1.4	103.6
1.6	103.7
1.9	104.6
2.2	105.1
2.5	105.0
2.8	106.3
2.9	107.6

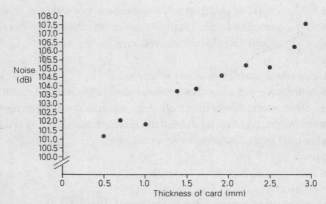

Figure 8.1.   Graph showing relationship between thickness of cardboard and noise in a banger

an important decision and there is a convention which makes life
simpler for us.

If, in an experiment or investigation, we vary one thing
(call it A) and look at its effect on another thing (B), then
A is the *independent variable* and B is the *dependent
variable*. When representing the data on a graph the independent variable always goes across while the dependent
variable goes up.

Alternatively, if we hope to use the behaviour of one thing to

predict the behaviour of another then the predictor goes across while
the thing to be predicted goes up.

B is called the dependent variable because we suspect, or hope,
that its behaviour depends on the behaviour of A which we can alter.
In many cases, however, we have no control over either variable and
so there is no clear choice for the horizontal axis. The decision then
becomes arbitrary or directed by some other factor such as scaling
on the graph paper.

Going back to the diagram again we can see that the points all lie
more or less in a straight line but are scattered around it. It is for this
reason that such a diagram is usually called a *scatter diagram*.

John Noble wanted to go further than just a general confirmation
of his feeling that the thicker the cardboard was made, the louder
the bang would become. He wanted to have some way of predicting
what volume of sound a given thickness would produce. To achieve
this he drew in a *line of best fit* on the scatter diagram. This was the
straight line which appeared to fit the set of points best.

There are sophisticated techniques available which would enable
us to calculate where this line should go (see *Statistics In Action*,
Peter Sprent, pp. 136 ff., Pelican, Harmondsworth 1977). The mathematics is quite difficult so I am not going to pursue this technique
further here. Essentially we shall draw in the line of best fit by eye
so that it looks as though it represents the trend of the data.

There is, however, one trick from the advanced theory which we
can use to make our line more accurate. It turns out that for a set of
points like those in Figure 8.1, the line of best fit will *always* pass
through the point $(\bar{x}, \bar{y})$ where $\bar{x}$ is the mean of the horizontal values
and $\bar{y}$ is the mean of the vertical values.

In John's case the mean thickness for his ten samples was 1.75mm
and the mean volume of noise was 104.08dB. If we plot this on our
diagram we can use it to draw in the line of best fit as shown in
Figure 8.2.

Having done this, John now had two options. He could use the
graph simply to read off any values he might be interested in. For
example, to produce a bang of 103dB, the graph told him that he
would need to make the tube at least 1.29mm thick.

A more powerful option would be to find the equation of the line
of best fit and to use this to calculate any other required values. This

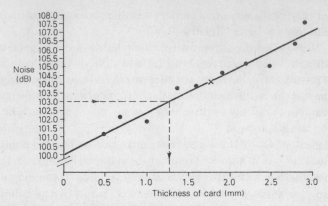

Figure 8.2.   *Graph showing relationship between thickness of card-board and noise in a banger*

calculation is dealt with in the section entitled 'Equations of lines' at the end of the chapter.

Correlation

In our example above there was a fairly obvious and straightforward connection between the two variables. This is not always the case. Indeed, very many research experiments are designed to try to discover or disprove connections between things. The measure of the amount of connection between two variables is usually referred to as the *degree of correlation* between them.

This may be measured in various ways. These measurements produce what are called *coefficients of correlation* and, regardless of the method of calculation, they all produce a value between −1 and +1. We are going to look at two differing coefficients – the Spearman Rank Correlation Coefficient and the Pearson Product-Moment Correlation Coefficient.

Do not be put off by these rather formidable titles. They are not really that difficult – certainly not the Spearman coefficient anyway. They are both named after their inventors as is the case with many of the measurements and distributions developed in the field of Statistics.

Before looking at these two, however, let us look at some of the different types of scatter diagrams which we might come across. Figure 8.3 illustrates the sort of patterns which may emerge depending on the degree of correlation, or lack of it, between the two variables.

I have here introduced the idea of positive and negative correlation. Positive correlation implies that as one variable gets bigger we can expect the other to do so as well. Negative correlation tells us that as one variable gets bigger we can expect the other one to get smaller and vice versa. The strength of the correlation as measured by the correlation coefficient tells us how much faith to put in our expectation.

So in our firework example, where the results lay pretty close to a straight line going up from left to right, we would say that there is a strong positive correlation between the thickness of the cardboard and the volume of the bang.

Spearman's Rank Correlation Coefficient

We now look at a more formal mathematical way of measuring the degree of correlation.

Researchers at the Normalia Institute for Audio Research are testing whether hearing is affected by age. They test a group of ten people of differing ages and find out the highest frequency that they can hear. The results are set out in Table 8.2. These figures lead to the scatter diagram shown as Figure 8.4 which indicates a reasonably strong negative correlation – the older you get the less you can hear. To quantify this relationship we first *rank* the ten individuals according to their age (place them in order) and then do the same based on their hearing. We then go on to process these ranks (Table 8.3).

While the calculation will not really be affected it is usual to assign the ranks in order of size. So the highest value of the variable is ranked 1 and the lowest 10. We could just as easily and as correctly

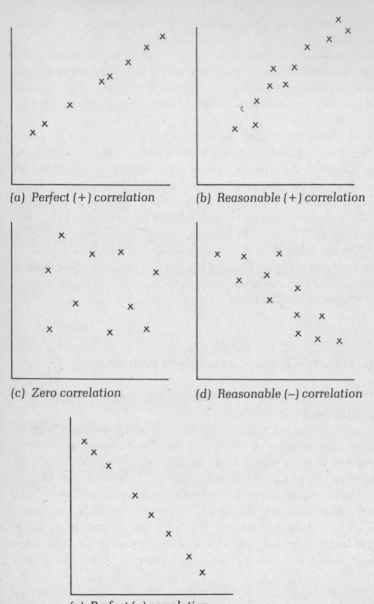

(a) Perfect (+) correlation

(b) Reasonable (+) correlation

(c) Zero correlation

(d) Reasonable (−) correlation

(e) Perfect (−) correlation

Figure 8.3.

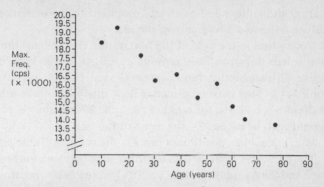

Figure 8.4. *Diagram showing relationship between hearing and age*

Table 8.2

Person	Age (years)	Maximum frequency (× 1000 cycles per sec. [cps])
A	9	18.4
B	15	19.3
C	24	17.6
D	30	16.2
E	38	16.5
F	46	15.3
G	53	16.2
H	60	14.8
I	64	14.1
J	76	13.9

Table 8.3

Person	Age (years)	Maximum frequency (× 1000 cps)	Age rank	Frequency rank
A	9	18.4	10	2
B	15	19.3	9	1
C	24	17.6	8	3
D	30	16.2	7	5.5
E	38	16.5	6	4
F	46	15.3	5	7
G	53	16.2	4	5.5
H	60	14.8	3	8
I	64	14.1	2	9
J	76	13.9	1	10

start at the bottom and work up, provided that we are consistent between the two sets of measurements.

Notice that in the case of the Maximum Frequencies there were two values the same. They provide us with what are known as *tied ranks*. This means that they share ranks 5 and 6 and so get the average rank – 5.5. This is true no matter how many ties there are. If five individuals had tied for ranks, 5, 6, 7, 8, 9 for instance, then they would each be ranked 7 – the mean of the tied ranks.

We are now in a position to process these ranks (Table 8.4).

The first extra column 'd' is the difference between the Frequency rank and the Age rank. $2 - 10 = -8$, etc. The second extra column contains the squares of these values to get around the old zero-total problem. We add up the last column to get Σd^2. Σ (the Greek capital 'sigma') is the symbol used for adding up a series of terms.

Table 8.4

Person	Age (years)	Maximum frequency ($\times 1000$ cps)	Age rank	Frequency rank	d	d^2
A	9	18.4	10	2	−8	64
B	15	19.3	9	1	−8	64
C	24	17.6	8	3	−5	25
D	30	16.2	7	5.5	−1.5	2.25
E	38	16.5	6	4	−2	4
F	46	15.3	5	7	+2	4
G	53	16.2	4	5.5	+1.5	2.25
H	60	14.8	3	8	+5	25
I	64	14.1	2	9	+7	49
J	76	13.9	1	10	+9	81
					Total Σd^2	320.5

The calculation of Spearman's Rank Correlation Coefficient, ρ (Greek letter 'rho'), is then

Number in sample $n = 10$

$$\rho = 1 - \frac{6\Sigma d^2}{n(n^2 - 1)}$$

$$= 1 - \frac{6 \times 320.5}{10(10 \times 10 - 1)}$$

$$= 1 - \frac{1923}{10 \times 99} = 1 - 1.942$$

$$= -0.942$$

This result is very close to -1 and so indicates a very strong negative correlation.

There are available further refinements which would enable us to say more about the value of ρ that we have obtained. This theory is an extension of probability theory and is well covered in *Quick Statistics* by Peter Sprent (Penguin, Harmondsworth, 1981), pp. 212ff. For our purposes it suffices to say that values for ρ of 0.9–1 are usually termed very strong, 0.7–0.9 strong, 0.5–0.7 moderate. This applies whether ρ is positive or negative. These values are, however, affected by the number of pairs of values we are dealing with.

Generally, the bigger the sample, the more scope there is for disagreement in ranking and so we must expect values of ρ nearer to 0. Thus a value for ρ of 0.534 when $n = 20$ has the same significance as a value of 0.943 when $n = 6$.

Exercise 8.1

1. A group of six people were weighed and had their heights measured. These were the results:

Person	A	B	C	D	E	F
Weight (kg)	56	67	62	67	51	73
Height (cm)	152	176	167	168	154	181

Calculate Spearman's Rank Correlation Coefficient for these figures.

2. The Normalia Film Corporation (NFC) produced five films in its first year. Their production costs and the number of people who paid to see them in their first six months of release are recorded in Table 8.5. Calculate Spearman's Rank Correlation Coefficient for these figures and explain what it ought to tell the NFC about film finance.

Table 8.5

Film	Cost N$1 000 000	Customers × 10 000
Flight to Fancy	5	76
Kali Kills Again	32	47
Love in the Ruins	24	61
Divine Endurance	9	84
The Trojan Way	17	63

3. In their end-of-year examinations the following marks were obtained by six pupils:

	Alison	Brian	Claude	Diana	Ethel	Franklin
Maths	45	60	32	81	45	76
Physics	64	64	40	78	56	92

Calculate Spearman's Rank Correlation Coefficient for these figures. Comment on your results.

4. Galactic Ice-creams are just about to launch a new range of ice-creams. Part of their research involved asking people to rank six new products according to taste, texture and visual appeal. Table 8.6 shows the results for one person who took part.

Table 8.6

Ice-cream type	Ranked by		
	Taste	Texture	Visual appeal
A	1	4	6
B	2	3	2
C	4	6	5
D	6	5	3
E	3	1	1
F	5	2	4

Calculate Spearman's Rank Correlation Coefficient for 'Taste' against 'Texture', for 'Texture' against 'Visual appeal' and for 'Visual appeal' against 'Taste'. Does there appear to be any connection between these attributes?

One of the big attractions of rank correlation is that the arithmetic is fairly straightforward which compensates for some loss of sophistication. Another point in its favour is the fact that it can be used to compare attributes or properties which cannot easily be given a numerical value. Generally these are aesthetic considerations such as taste, colour, appeal and beauty. Because of this it is very useful as a technique for market research when launching a new product. (For example: arrange these ten bottles of wine in order of preference for (a) taste, (b) smell.)

Numerical measures of properties such as weight, length, amount of time, number of atoms are called *quantitative variables* because they measure *quantities*. When processed they provide us with *parameters* with which to measure and predict behaviour.

Aesthetic measures such as those outlined above are called *qualitative variables* as they deal with *qualities* rather than quantities. They do not provide us with parameters. Methods of processing this kind of information are usually referred to as *non-parametric methods* and are becoming increasingly popular as more people become aware of their power and ease of application.

Another non-parametric statistic is Kendall's Coefficient of Rank Correlation. This takes slightly longer to calculate than Spearman's and so I leave it to Sprent (*Quick Statistics*, p. 214) to explain this. While the two results may be numerically different, there is usually not much difference in the interpretation of their significance.

Pearson's Product-moment Correlation Coefficient

If you are really hooked on parametric methods – because they are easier to program on a computer, for example – then there is a measure of correlation which is calculated from raw numerical data. It is called Pearson's Product-moment Correlation Coefficient and the formula for calculating it is

$$r = \frac{S_{xy}}{S_x S_y}$$

Here S_x and S_y are the standard deviations of our two variables x and

y. S_{xy} is called the *covariance* of x and y and is a measure of the amount they are linked. Its formula is similar in form to that for straight variance.

So

$$\text{Var}(x) = \frac{\Sigma(x_i - \bar{x})^2}{n} = \frac{\Sigma x_i^2}{n} - (\bar{x})^2$$

S_{xy} = the mean product of deviations =

$$\frac{\Sigma(x_i - \bar{x})\Sigma(y_i - \bar{y})}{n} = \frac{\Sigma x_i y_i}{n} - (\bar{x}\bar{y}).$$

It is this last form which makes the calculation easiest. If we go back to our original *bivariate data* (two variables associated with one thing) in Table 8.2, the calculation for it of Pearson's Product-moment Calculation Coefficient can best be set out as in Table 8.7.

Table 8.7

	x	y	x^2	y^2	xy
	9	18.4	81	338.56	165.6
	15	19.3	225	372.49	289.5
	24	17.6	576	309.76	422.4
	30	16.2	900	262.44	486.0
	38	16.5	1444	272.25	627.0
	46	15.3	2116	234.09	703.8
	53	16.2	2809	262.44	858.6
	60	14.8	3600	219.04	888.0
	64	14.1	4096	198.81	902.4
	76	13.9	5776	193.21	1056.4
Totals	415	162.3	21623	2663.09	6399.7

$\bar{x} = 415/10 = 41.5$

$\bar{y} = 162.3/10 = 16.23$

$S_x = \sqrt{(21623/10 - 41.5 \times 41.5)} = 20.98$

$S_y = \sqrt{(2663.09/10 - 16.23 \times 16.23)} = 1.70$

$S_{xy} = 6399.7/10 - 41.5 \times 16.23 = -33.575$

$$r = \frac{S_{xy}}{S_x \times S_y} = \frac{-33.575}{20.98 \times 1.70} = -0.941$$

Interpretation of this coefficient is much the same as previously discussed. Notice how close the value for this coefficient ($r = -0.941$) is to the Spearman Rank Correlation Coefficient. While this is not always going to be so, the two statistics will, when correctly interpreted, lead to essentially the same conclusions about the significance of any correlation. Before starting to use them, then, ponder on the effort involved in each calculation. As I said, unless you have a computer or some other pressing reason, the Rank Correlation Coefficients have a lot going for them.

Equations of lines

I offer here the mere mechanics of how to find the equation of a line. Explanations of the principles can be found in any elementary Mathematics textbook.

1. Choose two points which lie on the line. Try to make them so that the numbers are easy to work with. Suppose that we want to find the equation of the line passing through $(2, 4)$ and $(5, 13)$.

   All equations of straight lines can be written in the form $y = mx + c$ where m and c are two constants and x and y are the two variables. What we have to do is to find m and c.

2. To find m we work out the *gradient* of the line. This is the increase in y divided by the increase in x. So for our two points

   $$m = (13 - 4)/(5 - 2) = 9/3 = 3.$$

3. We now know that the equation of our line is $y = 3x + c$. To find c we use one of the points that we know and put its values into the equation. Thus

   $$4 = 3 \times 2 + c$$
   $$4 = 6 + c$$
   $$c = -2.$$

   Our equation is therefore $y = 3x - 2$.

4. Always check the answer by using the other point. In this case,

$$13 = 3 \times 5 - 2.$$

This is correct and so our equation is correct.

There is one case which can cause confusion. This occurs when y gets smaller as x gets bigger. In this case m is negative. For example, the gradient of the line joining (1,8) and (4,2) is

$$m = (2 - 8)/(4 - 1) = -6/3 = -2.$$

Exercise 8.2

1. Find the equations of the lines joining these pairs of points.

(a) (2,5) and (4,9).   (b) (3,14) and (7,42).

(c) (151,63) and (160,81).   (d) (240,363) and (280,283).

(e) (2,7) and (3,5).   (f) (4,19) and (7,12).

9 Time Series and Moving Averages

'*After seasonal adjustment the Normalian Trade Figures show an upward turn.*'

'*The weather over the last twenty years has got steadily colder.*'

'*New car sales have been declining in recent years.*'

These quotes are examples of a major application of statistics in everyday usage. They refer to patterns in what are called *time series*.

As an example of a time series and of some of the processing which we can apply to one, let us look at the goings on in the Mann-Whitney High School in Normalia. In recent years there has been a deterioration in the examination results obtained by fifth years. This has been especially pronounced with the Statistics results.

Feeling rather worried about this, Dawn Adams, the Head of the Statistics Department, decides that a major contributory factor is the worsening attendance record of fifth-year pupils. To back up her case she analyses attendance figures for the previous five years and comes up with Table 9.1. This is an example of a time series. Quite

Table 9.1. *Mean attendance rate (per cent) of fifth-year pupils over the last five years*

	Term		
Year	1	2	3
1	92.84	85.41	89.90
2	91.09	85.13	88.96
3	90.72	84.78	87.59
4	89.34	83.62	86.72
5	89.00	82.28	85.99

simply it is a set of values of a variable which have been recorded at regular intervals over a period of time.

If we look at the figures we see that they go up and down each year (predictably worst in the winter) and that there does seem to be a slight worsening over the five-year period. We can see this more clearly on a graph, Figure 9.1.

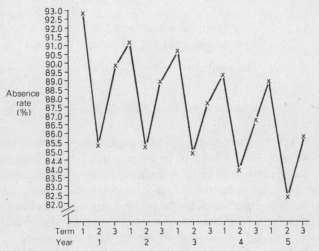

(*Note:* I have deliberately drawn this graph with a broken vertical scale to illustrate my theme here. This is a tactic which should only be employed with caution and moderation – unless your name is Gus.)

Figure 9.1.   *Graph showing percentage attendance by fifth years over the last five years*

I have chosen to join the points up with straight lines. This is intended to make patterns and trends easier to spot but does not mean that we can read off in-between values.

Moving averages

We can now see that our impression from the numbers is visually proven. What Dawn wants, however, is a numerical demonstration of this. To this end she calculates a set of *three-point moving averages* for the series. This is best explained by actually looking at her calculations.

Average 1 = Average of first, second and third values

$$= \frac{92.84 + 85.41 + 89.90}{3} = 89.38$$

Average 2 = Average of second, third and fourth values

$$= \frac{85.41 + 89.90 + 91.09}{3} = 88.80$$

Average 3 = Average of third, fourth and fifth values

$$= \frac{89.90 + 91.09 + 85.13}{3} = 88.71$$

Average 4 = ...
$$= 88.39$$

Averages 5 to 12 = 88.27, 88.15, 87.70, 87.24, 86.85, 86.56, 86.45, 86.00.

Average 13 = Average of thirteenth, fourteenth and fifteenth values

$$= \frac{89.00 + 82.28 + 85.99}{3} = 85.76$$

It is now clear that moving averages are so called because they are a set of averages which move through the main set of data. Their use is to iron out cyclical variation and to indicate the general *trend* of a time series. These numbers decline gradually and indicate that the general trend for the time series is, indeed, downwards. If we plot the moving averages on the original graph and join them up we get Figure 9.2.

Notice in Figure 9.2 that the first moving average gets plotted against term 2, that is, the middle of the averaging period. The second gets plotted against term 3 because this is the middle of 2, 3, 4 and so on until the last value (the thirteenth which gets plotted against term 14 – the middle of 13, 14, 15).

Dawn is quite happy with these results and shows them to her

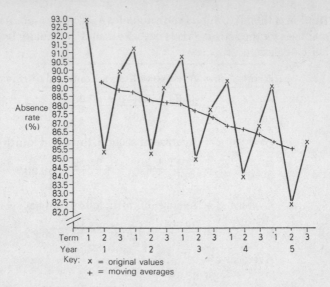

Key: x = original values
     + = moving averages

Figure 9.2.   *Graph showing percentage attendance by fifth years over the last five years*

headmaster who, impressed by the graph, feels less put out with her and agrees to try to do something about the attendance rates.

That was a fairly simple example. We must now look at a more complicated one and try to answer a typical question which is asked by production managers. We will start with some data (Table 9.2).

The manager wants to know the likely demand for January and February of the current year. This is necessary so that output can match demand without a shortfall or an overrun. The first step in answering this question is to produce a graph (Figure 9.3) to give an impression of what is happening.

There is, as one might expect, a fairly definite *seasonal variation*. More fuel is bought in winter than in summer. To try to determine the trend we are going to work out a set of moving averages. As before we shall make these three-point moving averages.

$$\text{Average 1} = \frac{46.8 + 47.0 + 45.5}{3} = 46.43$$

Table 9.2.   *Sales of fuel oil by the Normalia Oil*
Company over the last three years
(× 100 000 gals)

	Year 1	Year 2	Year 3
Jan.	46.8	48.0	51.3
Feb.	47.0	49.5	50.9
Mar.	45.5	47.7	49.5
Apr.	44.2	44.9	44.8
May	36.7	40.0	42.2
Jun.	33.8	39.3	39.2
Jul.	33.3	36.0	37.8
Aug.	34.2	36.0	38.0
Sep.	34.8	38.1	39.4
Oct.	37.8	42.7	44.5
Nov.	41.3	46.3	49.9
Dec.	46.5	50.2	49.6

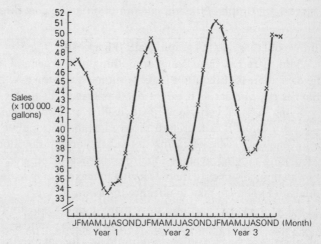

Figure 9.3.   *Graph of fuel oil consumption in the past three years*

To save space, the rest of the moving averages are, in order,

45.57, 42.13, 38.23, 34.60, 33.77, 34.10, 35.60, 37.97,
41.87, 45.27, 48.00, 48.40, 47.37, 44.20, 41.40, 38.43,
37.10, 36.70, 38.93, 42.37, 46.40, 49.13, 50.80, 50.57,
48.37, 45.50, 42.07, 39.73, 38.33, 38.40, 40.63, 44.60,
49.60,

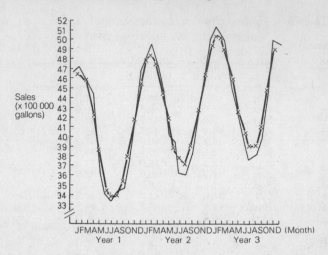

Figure 9.4.   *Graph of fuel oil consumption in the past three years*

and we put these on the graph again (Figure 9.4).

While there has been some smoothing of the general shape of the data, there is still quite a bit of variation along the new line. The trend is not as clear as it might be. The reason for this is that the overlying seasonal variation repeats itself every twelve months and not every three. It is therefore more sensible to calculate twelve-point moving averages. This is an important decision when calculating moving averages. In the absence of other criteria or information, fortunately there is a rule to make it easy to decide the number of points to take in our moving average.

Look at the graph and find how many points there are from one peak (or trough) to the next. This is known as the *period* of the *cycle*. It is also the number of points we should take for our moving average if we want to eliminate most variation.

In our case the peak value in year one occurred in February while in the second year it also occurred in February, twelve points later. This shows that the pattern is repeating itself every twelve months and so we must calculate a series of twelve-point moving averages. Note that if we only had quarterly values for the fuel consumption then the period would be four and we would calculate four-point moving averages.

Here are the first two calculations:

Average 1 =

$$\frac{46.8+47.0+45.5+44.2+36.7+33.8+33.3+34.2+34.8+37.8+41.3+46.5}{12}$$

$$= 40.16$$

Average 2 =

$$\frac{47.0+45.5+44.2+36.7+33.8+33.3+34.2+34.8+37.8+41.3+46.5+48.0}{12}$$

$$= 40.26$$

Even with a calculator, if we go on doing this it is going to be very tedious and time-consuming working out the twenty-five values that we want. Fortuately there is a trick which makes life easier for us.

The key to the short cut can be seen by looking at the two calculations above. In the first we added up twelve numbers to get a total (T_1 to remember it by) and then divided T_1 by 12 to get the average. In the second calculation eleven of the numbers are the same. All that has happened is that we have taken a value off the front of T_1 and put another one on the end.

Thus

$T_1 =$
$$46.8+47.0+45.5+44.2+36.7+33.8+33.3+34.2+34.8+37.8+41.3+46.5$$
$$= 481.9$$

and the second total

$T_2 =$
$$47.0+45.5+44.2+36.7+33.8+33.3+34.2+34.8+37.8+41.3+46.5+48.0$$
$$= 483.1$$
$$= T_1 - 46.8 + 48.0$$

By a similar argument

$$T_3 = T_2 - 47.0 + 49.5$$
$$= 485.6$$

and

$$\text{moving average } 3 = 485.6/12 = 40.47$$

Putting all of this together gives us a calculation for the twenty-five moving averages which looks like this:

Average 1 =

$$\frac{46.8+47.0+45.5+44.2+36.7+33.8+33.3+34.2+34.8+37.8+41.3+46.5}{12}$$

$$= 481.9/12 = 40.16$$

Average2 = (481.9 − 46.8 + 48.0)/12 = 483.1/12 = 40.26
Average3 = (483.1 − 47.0 + 49.5)/12 = 485.6/12 = 40.47
Average4 = (485.6 − 45.5 + 47.7)/12 = 487.8/12 = 40.65
and so on.

Even this can be shortened to make it easier for us to use a calculator. There is no real need for us to bother with the totals at all after the first calculation. We can do it all with the moving averages themselves. So the best way for us to present the calculation is as follows.

Average 1 =

$$\frac{46.8+47.0+45.5+44.2+36.7+33.8+33.3+34.2+34.8+37.8+41.3+46.5}{12}$$

$$= 481.9/12 = 40.16$$

Average		Value
2	(48.0 − 46.8)/12 + 40.16 =	40.26
3	(49.5 − 47.0)/12 + 40.26 =	40.47
4	(47.7 − 45.5)/12 + 40.47 =	40.65
5	(44.9 − 44.2)/12 + 40.65 =	40.71
6		40.98
7		41.44
8		41.67
9		41.82
10		42.09
11		42.50
12		42.92
13		43.23
14		43.47
15		43.62
16		43.77
17		43.76

18	43.94
19	43.93
20	44.08
21	44.25
22	44.36
23	44.51
24	44.81
25	44.76

In all of these calculations we have worked to two decimal places. This has been to preserve accuracy through the lengthy chain of working. The final task is to correct our twenty-five values to one decimal place. We do this because the original data was correct to only one decimal place and so we cannot really justify working to any greater accuracy in our working out. This gives us

40.2, 40.3, 40.5, 40.7, 40.7, 41.0, 41.4, 41.7, 41.8, 42.1, 42.5, 42.9, 43.2, 43.5, 43.6, 43.8, 43.8, 43.9, 43.9, 44.1, 44.3, 44.4, 44.5, 44.8, 44.8.

We now want to plot these values on our original graph. There is, unfortunately, a slight complication. In our last example we worked out three-point moving averages. When we came to plot these we used the middle value as the amount to go across. So our first point was plotted against 2 – the middle of 1, 2, 3. Now, alas, we have calculated twelve-point moving averages and there is no one value in the middle.

To get around this we plot the first value against 6.5 (in between June and July) as this is half way through the set. The second average we plot between July and August, and so on.

Having done this in Figure 9.5, a fairly smooth pattern emerges. This pattern can be used to comment on the figures which we already have or, more importantly, to help us to predict the next few values in the time series. This process of *extrapolation* can be carried out with varying degrees of sophistication.

The simplest method is to use the set of moving averages to predict what the next one is going to be. We can do this by drawing in a line of best fit for the points we have, as seen in Chapter 8. If we extend this line a little we can read off the next moving average. In our example this gives us a predicted value for the twenty-sixth moving

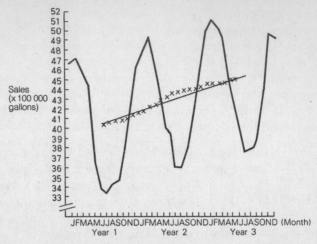

Figure 9.5. *Graph of fuel oil consumption in the past three years*

average of 45.4. We can now work backwards from this figure to find
a reasonable estimate for the January figure in Year 4.

$$\text{Average } 25 = 44.8$$
$$\text{Average } 26 = 45.4 = (? - 51.3)/12 + 44.8$$
$$\text{So } (? - 51.3)/12 = 45.4 - 44.8$$
$$? - 51.3 = 0.6 \times 12$$
$$? = 7.2 + 51.3$$
$$? = 58.5$$

We can reasonably predict, then, that the January figure will be about
$58.5 \times 100\,000$ gals. This process can be repeated as far into the future
as we want, although its reliability will, obviously, diminish the
further forward we go.

Exercise 9.1

1. Construct an accurate graph for the figures in Table 9.2.
   Superimpose on it the set of moving averages. Using the
   techniques of Chapter 8, draw a line of best fit through
   them. Hence work out predicted values for February and
   March of Year 4.

We must be careful to remember in the midst of all these seemingly accurate calculations that all we are doing is extending an apparent pattern. The whole series may be about to change dramatically due to some external and upredictable influence. This is likely in our example to be a Gulf war or sudden world overproduction affecting the price of the basic product and hence influencing demand.

It is also worthy of note that, while the projected figures are going to be reasonably accurate, they are unlikely to be exact. This can be seen from the behaviour of our moving averages which do not go up in an exact straight line. This is because, as well as the underlying *trend* and the superimposed *seasonal variation*, there is usually an overlying *random* (or unpredictable) *variation* which explains why the moving averages graph is rarely an exact straight line. The random variation will, one hopes, be small in comparison to the first two influences on the series. If it gets too large then our whole process of prediction becomes useless. Moving averages only iron out regular seasonal variations and are influenced by any random variations on top of these.

10 Collecting Data

GIGO (Garbage In, Garbage Out) is an acronym which applies equally well to Statistics as it does to its original application, computing. Gus understands this perfectly and is not above using it to his own advantage. It is also known as blinding them with science or, more famously, 'Lies, Damned Lies and Statistics.'

This chapter looks at one aspect of collecting data prior to statistical processing. A vast amount of research – sociological, economic and market – is done by conducting surveys of one form or another. There are several large organizations whose sole function is to conduct these surveys for other people. It is to the design and implementation of these surveys that we must now apply ourselves. Some of the points raised here will carry over into other fields where statistical data is gathered such as experimental psychology or particle physics.

Surveys

Very often we want to make general statements about a large *population*. This may be an attempt to predict the average life expectancy of a Wonderlite light bulb or to predict which way an electorate will vote. In most cases it is impractical to test the whole of the population and so we must satisfy ourselves with conducting a survey of a *sample* from the *parent population*. It should be understood that here 'population' is a precise statistical term which refers to everyone or everything under examination. It originates from the obvious meaning but could now be used about all the TV sets made by a company or all the wheat grown in Normalia.

The behaviour of the sample is used to predict the behaviour of the parent population; it is hoped quite closely. Indeed, estimation of the degree of error in such approximations is one of the fine arts and most powerful applications of Statistics.

Carrying out a survey

This can be broken down into eight steps.

1. Decide what you want to find out.
2. Decide who you are going to ask.
3. Decide what form the questioning is going to take.
4. Prepare a questionnaire.
5. Conduct a pilot survey and amend the questionnaire in the light of its results.
6. Conduct the actual survey.
7. Process the results.
8. Comment on them.

We have spent quite a lot of time on steps seven and eight, so we shall now look at the first six in turn.

What do we want to find out?

This seems so obvious a question that it is easy to dismiss it, but it is probably the most important of all. Good experimental design and a clear statement of the objectives will save many heartaches later.

A survey may well involve a lot of time, effort and money, so it is worth clarifying at the start what it is purporting to find out.

Take the case of the Normalia Tea and Coffee Company. They were about to launch a new brand of quality tea and were concerned that the packaging should be correct. They designed six different boxes for it, ranging from the staid and traditional in a nice shade of dark green to the flash and trendy in silver and yellow. A market-research organization went out with these and asked 1000 people which one they liked best. 843 voted for the silver and yellow. The NTCC duly launched the tea in this box and sales were disastrous.

In the aftermath amid the acrimony it turned out that the survey had neglected to find out anything about the drinking habits of those interviewed. If they had done so they would have discovered that the regular tea drinkers in the sample all favoured dark green . . .

Who are we going to survey?

In that example 1000 people were surveyed of whom only a minority were interested in tea. It is very important when designing a survey that we choose the *sample* to be investigated in such a way as to minimize *bias* or misrepresentation. We must remember at all times that the sample is supposed to represent the parent population and if it does not, the results of our survey will be of no use to us.

To this end there are various ways to select our sample. These can be separated into three types.

1. Purposive sampling.
2. Random sampling.
3. A mixture of 1 and 2.

Purposive sampling

This means that we choose our sample with some underlying theme or purpose. We may go out and interview everyone we meet who has grey hair or who is wearing a brown overcoat. The Normalia Tea and Coffee Company would have been well advised to employ some form of purposive sampling since they were really only interested in the views of tea drinkers.

Purposive sampling can obviously be a source of bias. If, in a survey of national income in the United Kingdom, the sample was selected deliberately from Belfast or from Surrey it would not be a fair reflection of the overall national picture. This is where Gus gets his edge. Depending on the point he wants to make, he will choose his sample with care.

Thus, if employed by a poverty action group, he would choose Belfast and come up with the line – perfectly valid and perfectly meaningless as a basis for generalizing – that of 1000 people surveyed in the United Kingdom the net average income was £x (where x is a small number). The same statement based on 1000 people from selected parts of Surrey could present a very different kind of picture. It might well be the one used by the political party in power. Beware the results of surveys where the nature of the sample is not quoted!

Random sampling

Random sampling eliminates in-built bias by ensuring that no member of the population – be it car, light bulb or person – is more likely to be chosen than any other. Indeed that gives us the formal definition of a random sample.

> A sample of a population is said to be randomly chosen if every member of the population has the same probability of being included in the sample.

While this is easily stated it is not always too easy to achieve in practice. A frequently used tool in selecting such a sample is a table of random numbers. Here is an extract from such a sample.

> 3968, 3051, 8534, 8076, 6224, 0136, 0476, 6020, 4131,
> 3539, 0828, 6091, 0617, 0333, 8524, 0318, 1894, 3059,
> 8155, 0167, 5508, 7781, 0290, 7085, 9033, 0461, 9306.

In actual fact there is a lot of heated discussion about what is meant by random numbers and how to generate them. For our purposes a table such as this is more than adequate without getting embroiled in the philosophy behind it.

Suppose that a customs official at Normalia docks is confronted with 5000 cardboard boxes purporting to contain cheap wine. She is not going to open them all but she wants an idea of what the whole batch is like. She decides to open 100 boxes, one out of every fifty. She could just open every fiftieth box in turn but such predictable behaviour would be playing into the hands of would-be smugglers. Instead she imagines that the boxes are numbered from '1' to '5000' and then she consults a table of random numbers. Starting anywhere in the table she reads off 100 four-figure numbers ignoring any over 5000. These are the boxes which she opens.

There will be no bias in her choice of boxes and so she is more likely to catch any smuggling. If left to 'free' choice she would run the risk of being manipulated by a smuggler in much the same way as a skilled magician can manipulate a person into choosing a selected card or cards.

Multi-stage sampling

Unfortunately in our example it is going to be rather tedious for the sampler to select the 3914th and then the 1821st box. It would also be rather expensive to interview someone in Stornoway, someone in Plymouth, someone in Cardiff...

To get around this problem and to get the best of both worlds when choosing our sample we often mix the two ideas. Going back to our customs official, life would be easier for her if she split the 5000 boxes up into lots of 50 giving 100 batches. From each of these she will choose one box by selecting a random number in the range 1–50. This is an example of *multi-stage sampling* because the sampling is done in several stages. It is a widely used technique to minimize cost and effort.

A fuller example can be found if we look at a national survey of voters in the UK. If we assume that we want a sample size of 3000 we could go to the collected electoral registers held in London and choose 3000 people at random. To then get to them would be very expensive and time-consuming. We can rule out conducting the survey by phone because this would immediately bias our sample by excluding the fairly large section of the electorate who do not have telephones.

Such a survey really needs to be person to person. So, out of all the constituencies in the country we choose at random, using our tables, ten (say). This is the first stage. The second stage is to survey 300 voters from each of the selected constituencies.

We now have to decide how we are going to get our sample of 300 from each target constituency. Again, we might do a further stage by choosing five wards, and finally take 60 voters at random from each of them. In such a way we can ensure that bias has been avoided to the best of our ability.

Of course a random sample may end up being unrepresentative. The 3000 people chosen at random by our multi-stage method could turn out to be the only 3 000 people in the whole country who have decided to vote for the Raving Loony Party.

However 3 000 is a fairly large sample and we can be pretty confident that it is going to be representative. It is rare, in fact, for such surveys to involve more than 1 000 people.

Stratified sampling

Another stratagem employed by researchers to help to eliminate accidental bias in the sample is a technique called *stratified sampling*. In essence, the target population is divided into various *strata* or linked groups by some rule. Frequently this is socio-economic but it may be based on things as diverse as eye-colour or shoe size.

The first step is to identify within the population the differing strata we want. We then go out and ensure that our sample includes members of each stratum. The requirement may simply be that all strata are to be represented in the sample. More usually a *quota* is employed whereby a certain number have to come from each stratum. The size of the quota is determined either by the relative size of each stratum or, more frequently in the case of market research, the perceived relative importance of each stratum.

Thus, if we conducted a general survey into the spending habits of the Normalian population we might well use social classes A, B, etc. as our strata and ensure that our quotas reflected the relative numbers in each group on a national level.

If, on the other hand, we were a large multinational company looking to cash in on the booming leisure market, our quotas would be skewed in favour of the young and affluent as this would be the perceived potential market.

Types of survey

There are four principal methods of conducting a survey. To a certain extent they determine the size and nature of the sample to be examined and so cannot really be considered separately from the previous section. They are:

1. postal;
2. telephone;
3. face to face;
4. electronic.

A brief outline of the advantages and disadvantages will suffice.

Postal surveys

This is a cheap way of reaching a lot of people. In theory the whole of the population can be reached by this means. On the other hand, not very many of them will bother to reply unless there is some very strong incentive.

Only people of a certain type or those with an axe to grind tend to respond to postal surveys and so they are prone to bias.

Telephone surveys

Again, this is a relatively cheap survey method. The interviewer can reach a very large area quickly and without having to travel. A major drawback is that not all of the population can be easily reached over the telephone and so, depending on the type of sample we want, the survey may be unrepresentative.

Face-to-face surveys

These allow for a greater depth of questioning and greater detail in response. Unfortunately it is an expensive method, especially if we want to sample from a large area when time and travel costs will weigh heavily.

Electronic surveys

These are becoming increasingly popular as the equipment becomes cheaper and more sophisticated. Audience research figures for television are now gathered in this way, as are returns about record sales. The big advantage is that once the system is set up it is cheap to run and tireless in operation.

A disadvantage is its lack of flexibility. When researching viewing figures, the machinery can easily say for how many hours and at what times each channel was being relayed through the TV. Unfortunately it cannot yet register how many people, if any, are watching. Doubtless time will change that . . .

Preparing the questionnaire

The exact nature of the questionnaire will depend on all of the factors already discussed, but there are some underlying themes. All questions should be:

1. unambiguous;
2. not leading;
3. limited in the range of responses which can then be easily classified for processing;
4. clear and easily answered by the least intelligent person in the sample;
5. polite!

The nature of the questions is determined by the type of survey. In a postal survey where there will be no one to explain the questions it is vital that they should be easy to answer. With telephone or face-to-face methods this is not so important as the interviewer can help with any difficulties. Clarity should still, however, be a design aim. If the interviewer, unscripted, starts 'clarifying' questions she or he may introduce bias into the responses.

Conduct a trial run

This can save a lot of time and trouble. A few people are asked to respond to the questionnaire and, in the light of any difficulties, it is amended before use.

Conduct the actual survey

By now this should need no further explanation.

One final word of warning concerns the problem of non-respondents. It is tempting, but dangerous, to ignore them and to reduce the sample size. If 60 per cent of a targeted sample actually respond (this would be an extremely good response rate), should we be content to do nothing about those who did not participate?

In actual fact we do not really have any choice, but, when publishing our findings, we ought to include information about the response rate so that the reader can have a full picture of the results. Usually there is a linking theme or themes among non-respondents. By ignoring them, we are in danger of missing an entire stratum or strata.

Door-to-door surveys conducted at 11 a.m. will tend to miss those people with daytime jobs. People in the age range 16–25 are notoriously bad at helping with surveys. The interviewer may have approached only those people whom he or she liked the look of.

Surveys are prone to misrepresent the true nature of the parent population. Opinion polls before elections show this to be the case. However, if they are properly designed, implemented and interpreted they can be extremely useful in many areas.

11 Numbers, Calculators and Computers

Calculators and computers have liberated us from much of the drudgery of arithmetic. They make the processing of statistical data very much easier than it used to be. This ease of use and apparently complete accuracy can, however, be a pitfall as well as a blessing.

As part of a recent calculation, I was working out the mean of thirteen measurements. The total for the thirteen measurements was 49 giving a mean of $49/13 - 3.7692308$cm. All very straightforward on a calculator, but if I had left this as a final answer it would have been rather misleading. All measurements are made to a certain degree of accuracy and it would be a remarkable experiment that produced a measurement such as 3.7692308cm.

When a number is written using our decimal (counting in tens) notation, each digit has what is called *place value*. This means that its importance varies depending on its place in the number. So, in my result above, the digit 3 appears in two different places. In its first appearance, to the left of the decimal point, it stands for 3cm which is quite visible when drawn as a line.

On its second appearance it is very far down the pecking order of place value which reads 7 tenths, 6 hundredths, 9 thousandths, 2 ten-thousandths and 3 hundred-thousandths of a cm. 3/100000 of a cm is not a very great distance! In fact it is about 1/200th of the thickness of this paper.

This means that, in the context of the number 3.769 230 8, the first appearance of the digit 3 is a lot more *significant* or important than the second (100 000 times more significant in fact). This idea of *significant figures* (s.f.) is essential if we are going to make sensible use of a calculator or computer.

Most measurements which we are likely to come across will be relatively crude. When asked my weight I am going to say 67kg (if I am feeling doggedly metric) or 10st 7lb under more normal circumstances. I am not going to quote grams or ounces since these will vary dramatically depending on what I have been doing. It is usual when quoting measurements like these to give only the first few significant figures (one, two or three). Thus the height of Mount Everest is usually quoted as 29000ft rather than 29002 as the last 2 is not particularly significant in the context of the other 29000.

This leaves us with the problem of deciding on how many figures we are going to call significant and how we are going to amend our results to suit this decision. In the vast majority of measurements, three or four significant figures are the normal order of accuracy. We shall rarely, therefore, be justified in giving statistical measurements such as mean values or standard deviations to any greater degree of accuracy than this.

So how do we quote, correct to three s.f., a number which our calculator tells us is 3.7692308? It is tempting to just write down 3.76 as these are the three digits furthest to the left. While this is almost correct, it tells a slight lie. Our answer, 3.7692308, is, in fact, nearer to 3.77 than it is to 3.76. This means that 3.77 is a better representation of it correct to three s.f. than is 3.76. This process of rounding is very important.

£2400000 becomes £2000000 when quoted to one s.f. (or to the nearest million) while £2600000 becomes £3000000. The dividing line between *rounding down* and *rounding up* comes with £2500000 which is exactly half way between the two values. There is no especial reason to round it one way or the other. It would be perfectly justifiable to round it either up or down. Because of the possible ambiguity here, the convention is that we *always* round a five up. So £2500000 gets rounded to £3000000.

Exercise 11.1

1. Round these figures off correct to (a) three s.f. (b) two s.f. (c) one s.f.

   (1) 4246   (2) 53.845   (3) 467800   (4) 690.1
   (5) 2.0052   (6) 407.8   (7) 3.14159   (8) 93247121.

There are three potential difficulties associated with rounding and significant figures.

The first comes when we have a number like 44.6. Correct to two s.f. this becomes 45. If we then want to give it correct to one s.f. it is tempting now to write 50 since 45 gets rounded up. This would be wrong because the original number, 44.6, is closer to 40 than to 50. Always, when doing successive shortenings, refer back to the original number.

Another problem can arise if we want to round off a number like 9.7 correct to one s.f. This becomes 10 which looks rather odd since it is very dissimilar to the original number. This occurs again if we round 99.6 off to two s.f. giving 100. There is always a temptation to call this 99 instead but consistency tells us that, because the first digit to be dropped is a 6, we must round 99.6 up to the number above (100).

The third and most complicated problem comes if we have a measurement like 0.0004638km. Correct to three s.f. it is tempting to write 0.00 or 0.000. This, unfortunately, would lose all information about the original measurement other than that it was rather small. It would make comparisons with a distance like 0.000005371km (= 0.000) rather difficult.

Sticking with the idea of significant figures, as before, the most significant figure in 0.0004638 is the 4 since this tells us most about the size of the thing we are measuring. As a consequence we write our measurement as 0.000464km (three s.f.) and 0.000005371 becomes 0.00000537km (three s.f.).

We have to be careful, having ignored the leading zeroes, not to do the same with any which appear surrounded by other digits. This means that 0.030 42 = 0.030 4 (three s.f.) = 0.030 (two s.f.). Here the 0 between the 3 and the 4 is counted as significant unlike the two zeroes before the 3.

Exercise 11.2

1. Round these figures off correct to (a) three s.f., (b) two s.f., (c) one s.f.

   (1) 9.553   (2) 998.4   (3) 9834   (4) 0.9923
   (5) 64.83   (6) 0.001034   (7) 0.000004651
   (8) 0.00009847

Decimal places

Sometimes instead of quoting a number to three or four s.f. we abbreviate instead to a certain number of *decimal places* (d.p.) The principle of shortening is exactly the same as before but here we give a predetermined number of figures after the decimal point regardless of what is going on before the point.
So:

$$3.769\ 230\ 8 = 3.769\ 23 \quad \text{(fived.p.)}$$
$$= 3.769 \quad \text{(threed.p.)}$$
$$= 3.8 \quad \text{(oned.p.).}$$

A number such as 12.904 should be quoted as 12.90 (two d.p.) rather than 12.9 (two d.p.) since the 0 shows the degree of accuracy to which we are working and that we have not just made a mistake or forgotten to finish the number off. 12.90 could be anywhere between 12.895 and 12.905 while 12.9 could be between 12.85 and 12.95: a much wider range.

Note that in the case of decimal places a number like 0.000 53 = 0.00 (two d.p.) and the significant part vanishes.

Exercise 11.3

1. Write each of these numbers correct to (a) three d.p., (b) two d.p., (c) one d.p.

   (1) 3.14159   (2) 9.10051   (3) 0.6992
   (4) 0.471893   (5) 2.17163   (6) 412.3689

Errors

The margin within which our measurements always lie gives rise to possible error. If I measured a rectangular piece of paper and told you that it was 11 cm by 8 cm to the nearest cm, you would probably tell me that its area was $11 \times 8 = 88\text{cm}^2$. This is not necessarily true. Indeed, it is highly unlikely to be true. The area could be as little as

$$10.5 \times 7.5 = 78.75 \text{cm}^2$$

or as great as

$$11.4999 \ldots \times 8.4999 \ldots$$

which in this context is usually written as

$$11.5 \times 8.5 = 97.75 \text{cm}^2.$$

All that we can say for certain is that the area lies somewhere between these two values. If we quote the area as 88cm^2 then the maximum *absolute* error is $97.75 - 88 = 9.75 \text{cm}^2$. In comparison to the area of about 88cm^2 this is quite a lot. If, however, we measured the area of a field or even a room, and were only this much out, we would be very pleased. The size of the error has to be viewed in relation to the thing being measured.

This *relative* error is usually quoted as a percentage. In our case the percentage relative error is

$$\frac{9.75}{88} \times 100 = 11.1 \text{ per cent (three s.f.).}$$

This is quite high and might be a source of concern.

In all cases where we are processing figures we must be careful about the potential accumulation of error caused by rounding measurements and numbers off. Usually the errors tend to cancel each other out but we cannot guarantee this. Computer manufacturers are much troubled by this problem even though they count in twos and not in tens.

As an example of the way in which these errors act together, consider these three values which have been provided for a, b and c:

$$a = 4.6 \quad b = 3.7 \quad c = 1.4 \text{ (one d.p.).}$$

We are asked to find the greatest possible relative error in calculating the expression

$$x = \frac{a}{b - c}.$$

$$x = \frac{4.6}{3.7 - 1.4} = \frac{4.6}{2.3} = 2.$$

For the greatest value we want the top line to be as big as possible
and the bottom line to be as small as possible. The top line is easy:
$a = 4.65$. To make $b - c$ as small as possible we want b to be small
and c to be big ($b = 3.65$, $c = 1.45$). Putting this together we get the
greatest value for x

$$x = \frac{4.65}{3.65 - 1.45} = 2.11 \text{ (three s.f.)}$$

Similarly the least value is

$$x = \frac{4.55}{3.75 - 1.35} = 1.90 \text{ (three s.f.)}.$$

The greatest absolute error $= 2.11 - 2 = 0.11$

The greatest relative error $= \dfrac{0.11}{2} \times 100 = 5.5 \text{ per cent.}$

Exercise 11.4

1. Four machined pieces of steel were found to weigh

   15.0, 15.3, 15.7, 15.6 g (one d.p.).

   Find the greatest and least values for the total weight of
   the four specimens and hence for the mean weight of the
   four. Find the maximum possible absolute and relative
   error involved in the calculation.

2. $a = $ cm/sec., $b = 8$ cm/sec., $c = 3$ cm/sec. (all correct to the
   nearest whole number).
      Find the greatest possible value of the expression

   $b^2 - 4ac$.

   Using the values for a, b and c which made this a
   maximum, find the relative error in calculating

   $$x = \frac{-b + \sqrt{(b^2 - 4ac)}}{2a}.$$

Recommended Further Reading

Readers will have noticed references to various other books throughout the text. To save working back through the text to follow up on points or areas in which you are interested, these books are listed below. They are all worth reading in their own right.

Darrell Huff, *How To Lie With Statistics*, Pelican, Harmondsworth, 1981

M. J. Moroney, *Facts From Figures*, Pelican, Harmondsworth, various editions and reprintings, 1951–84

Derek Rowntree, *Statistics Without Tears*, Penguin, Harmondsworth, 1981

Peter Sprent, *Quick Statistics*, Penguin, Harmondsworth, 1981

Peter Sprent, *Statistics in Action*, Pelican, Harmondsworth, 1977

K. A. Yeomans, *Introducing Statistics*, Penguin, Harmondsworth, 1968

Appendix: Some BASIC Programs

As promised, here are some computer programs to take the drudgery out of your statistical life. They are written in BASIC. Since this was first developed as a language it has grown and changed. Every machine now has its own dialect of it. This has made the task of writing these programs difficult. As far as it is possible I have used a standard subset of BASIC which seems to be shared by most home computers. This means that in a variety of ways they can be made more elegant depending on your machine. Feel free to do so.

Having said that, I want to use random numbers which are non-standard in implementation. I have decided to write the programs so that they will run on a Sinclair Spectrum. To adapt them for other machines the only change which is necessary is to replace 'RND', wherever it appears, by the command which, on your machine, will produce a random number in the range 0 to 1. You can also, if your machine allows it, leave out 'LET' where it appears.

I have tried to make the programs self-explanatory in function. To this end they are liberally supplied with 'REM' statements. Leave these out if you cannot be bothered typing them in or if you want the program to run faster.

1. Tossing a coin

```
10  REM * PROGRAM TO SIMULATE TOSSING A COIN
20  PRINT "HOW MANY TOSSES DO YOU WANT";
30  INPUT N
40  LET H = 0
50  LET T = 0
60  REM * H = NUMBER OF HEADS
70  REM * T = NUMBER OF TAILS
80  REM * N = NUMBER OF TOSSES
```

```
 90  FOR X = 1 TO N
100  LET A = RND
110  REM * TOSS THE COIN
120  IF A < 0.5 THEN LET H = H + 1
130  REM * A HEAD
140  IF A > 0.5 THEN LET T = T + 1
150  REM * A TAIL
160  IF A = 0.5 THEN 100
170  REM * ON THE EDGE!
180  NEXT X
190  PRINT:PRINT"NUMBER OF HEADS = ";H
200  PRINT:PRINT"NUMBER OF TAILS = ";T
210  REM * OUTPUT THE RESULTS
```

2. Sequences of heads when tossing a coin

```
 10  REM * PROGRAM TO TOSS A COIN AND COUNT
 20  REM * THE LENGTHS OF SEQUENCES OF HEADS
 30  DIM L(15)
 40  REM * THIS COUNTS THE FREQUENCY FOR EACH
     LENGTH
 50  REM * 15 IS AN ARBITRARY NUMBER WHICH SHOULD
 60  REM * COVER MOST CASES
 70  LET C - 0
 80  REM * THIS COUNTS THE LENGTH OF A SEQUENCE
 90  PRINT"HOW MANY TOSSES DO YOU WANT";
100  INPUT N
110  REM * N = NUMBER OF TOSSES
120  FOR X = 1 TO N
130  LET A = RND
140  REM * TOSS THE COIN
150  IF A < 0.5 THEN GOSUB 1000
160  REM * PROCESS A HEAD
170  IF A > 0.5 THEN GOSUB 2000
180  REM * PROCESS A TAIL
190  IF A = 0.5 THEN 130
200  REM * ON ITS EDGE!
```

```
 210  NEXTX
 220  GOSUB2020
 230  REM * COUNT LAST SEQUENCE
 240  PRINT:PRINT"LENGTH OF SEQUENCE","FREQUENCY"
 250  FOR X = 1 TO 15
 260  PRINT X,L(X)
 270  REM * OUTPUT THE RESULTS
 280  NEXT X
 290  STOP
1000  REM * SUBROUTINE TO PROCESS A HEAD
1010  PRINT"H";
1020  LET C = C + 1
1030  REM * INCREMENT SEQUENCE COUNTER
1040  RETURN
2000  REM * SUBROUTINE TO PROCESS A TAIL
2010  PRINT"T";
2020  IF C = 0 THEN RETURN
2030  REM * NO SEQUENCE TO COUNT
2040  LET L(C) = L(C) + 1
2050  REM * INCREMENT FREQUENCY COUNTER
2060  LET C = 0
2070  REM * RESET SEQUENCE COUNTER
2080  RETURN
```

3. Tossing four coins

```
 10  REM * PROGRAM TO SIMULATE TOSSING 4 COINS
 20  REM * AND TO RECORD THE NUMBER OF HEADS EACH
     TIME
 30  DIM T(5)
 40  REM * THIS WILL STORE THE RUNNING TOTALS
 50  PRINT"HOW MANY TOSSES DO YOU WANT";
 60  INPUT N
 70  REM * N = NUMBER OF TOSSES
 80  FOR X = 1 TO N
 90  LET C = 0
100  REM * C COUNTS THE NUMBER OF HEADS
```

```
110  FOR Y = 1 TO 4
120  REM * TOSSING 1 COIN 4 TIMES
130  LET A = RND
140  REM * TOSS THE COIN
150  IF A < 0.5 THEN LET C = C + 1
160  REM * A HEAD SO INCREMENT COUNTER
170  NEXT Y
180  LET T(C+1) = T(C+1) + 1
190  REM * A QUIRK OF THE SPECTRUM!
200  NEXT X
210  PRINT:PRINT"NUMBER OF HEADS","FREQUENCY"
220  FOR X = 0 TO 4
230  PRINT X,T(X+1)
240  REM * OUTPUT THE RESULTS
250  NEXT X
```

A quirk of the Spectrum is that, unlike most machines, its arrays
start at T(1) and not T(0). To record the results, then, it is necessary
to record the frequency for 0 in T(1), the frequency for 1 in T(2) and
so on. If your machine allows T(0) then, for the sake of elegance,
delete the +1 inside the brackets in lines 180 and 230.

4. Tossing a die

```
 10  REM * PROGRAM TO SIMULATE TOSSING A DIE
 20  DIM T(6)
 30  REM * COUNTER FOR SCORES
 40  PRINT"HOW MANY THROWS DO YOU WANT";
 50  INPUT N
 60  REM * N = NUMBER OF TOSSES
 70  FOR X = 1 TO N
 80  LET A = INT(6 * RND) + 1
 90  REM * CREATE A RANDOM WHOLE NUMBER
100  REM * BETWEEN 1 AND 6
110  PRINT A + " ";
120  REM * DISPLAY RESULT
130  LET T(A) = T(A) + 1
```

```
140  REM * RECORD RESULT
150  NEXTX
160  PRINT:PRINT"SCORE","FREQUENCY"
170  FOR X = 1 TO 6
180  PRINT X,T(X)
190  REM * OUTPUT THE RESULTS
200  NEXT X
```

5. Tossing two dice

```
 10  REM * PROGRAM TO TOSS TWO DICE
 20  REM * AND RECORD THEIR TOTAL
 30  DIM T(12)
 40  REM * THIS WILL STORE THE RUNNING TOTALS
 50  REM * FOR EACH SCORE
 60  PRINT"HOW MANY THROWS DO YOU WANT";
 70  INPUT N
 80  REM * N = NUMBER OF THROWS
 90  FOR X = 1 TO N
100  LET A = INT(6 * RND) + 1
110  LET B = INT(6 * RND) + 1
120  REM * TOSS THE TWO DICE
130  LET S = A + B
140  REM * S = SUM OF TWO SCORES
150  LET T(S) = T(S) + 1
160  REM * RECORD THE SCORE
170  PRINT S + " ";
180  REM * DISPLAY THE SCORE
190  NEXT X
200  PRINT:PRINT"SCORE","FREQUENCY"
210  FOR X = 2 TO 12
220  PRINT X,T(X)
230  REM * OUTPUT THE RESULTS
240  NEXT X
```

6. Processing data

```
  10 REM * PROGRAM TO PROCESS STATISTICS
  20 REM * AND TO PRODUCE MEASURES OF THEM
  30 REM * INPUT SECTION
  40 GOSUB1000
  50 REM * SORT THE DATA
  60 GOSUB2000
  70 REM * CALCULATE THE STATISTICS
  80 GOSUB3000
  90 REM * OUTPUT THE RESULTS
 100 GOSUB4000
 110 STOP
1000 REM * SUBROUTINE TO INPUT THE DATA
1010 PRINT"YOU WILL NEED A ROGUE VALUE"
1020 PRINT"TO TERMINATE THE DATA."
1030 PRINT"WHAT IS IT TO BE";
1040 INPUT R
1050 PRINT:PRINT"LAST VALUE MUST BE ";R:PRINT
1060 DIM D(100)
1070 REM * D( ) WILL HOLD THE DATA
1080 REM * UP TO 100 ITEMS
1090 LET N = 0
1100 REM * N = NUMBER OF DATA ITEMS
1110 PRINT"INPUT DATA ITEM NUMBER ";N+1;
1120 INPUT Z
1130 REM * Z IS DATA TO BE PROCESSED
1140 IF Z = R THEN GOTO 1300
1150 REM * TEST FOR END OF DATA
1160 LET N = N + 1
1170 REM * INCREMENT DATA COUNTER
1180 LET D(N) = Z
1190 REM * NOT LAST ITEM SO PUT INTO ARRAY
1200 REM * AND THEN GO BACK FOR NEXT ITEM
1210 GOTO1110
1300 REM * ALL THE DATA IS NOW IN
1310 REM * SO NOW CHECK IT
1320 PRINT"DO YOU WANT TO REVIEW THE DATA (Y OR N)?"
```

```
1330 INPUT A$
1340 REM * RESPONSE SHOULD BE Y OR N
1350 IF A$ = "N" THEN RETURN
1360 REM * INPUT OK
1370 IF A$ <> "Y" THEN PRINT "Y OR N ONLY
     PLEASE!":GOTO1330
1380 REM * ACCEPT CORRECT RESPONSE ONLY
1390 REM * YES SO REVIEW THE DATA
1400 PRINT:PRINT"YOU WILL SEE THE DATA ITEMS AND
     THEIR NUMBERS"
1410 PRINT:PRINT"IF ONE IS WRONG THEN PRESS C
     (CHANGE)"
1420 PRINT:PRINT"IF IT IS ACCEPTABLE PRESS A"
1430 FOR X = 1 TO N
1440 PRINT X,D(X)
1450 REM * DISPLAY DATA
1460 LET A$ = INKEY$
1470 IF A$ <> "C" and A$ <> "A" THEN GOTO 1460
1480 REM * CORRECT RESPONSE ONLY
1490 IF A$ = "A" THEN GOTO 1540
1500 REM ACCEPT DATA AND MOVE ON
1510 PRINT"NEW VALUE = ";
1520 INPUT D(X):PRINT D(X)
1530 REM * CHANGE THIS ITEM
1540 NEXT X
1550 PRINT:PRINT
1600 REM * ALL DATA SHOULD NOW BE CORRECT SO
1610 RETURN
2000 REM * SUBROUTINE TO SORT THE DATA
2010 REM * BUBBLE SORT
2020 FOR X = N TO 2 STEP -1
2030 REM * PUT LARGEST VALUE AT THE END
2040 REM * AND WORK BACKWARDS
2050 FOR Y = 1 TO X-1
2060 REM * WORK THROUGH THE ARRAY
2070 IF D(Y) < D(Y+1) THEN GOTO 2150
2080 REM * IN CORRECT ORDER SO DON'T SWAP
2090 LET T = D(Y)
```

```
2100  LET D(Y) = D(Y+1)
2110  LET D(Y+1) = T
2120  REM * SWAP ROUTINE SO Y AND Y+1
2130  REM * ARE NOW IN CORRECT ORDER SO
2140  REM * MOVE ON DOWN ARRAY
2150  NEXT Y
2160  REM * LARGEST ITEM NOW MOVED TO POSITION SO
2170  NEXT X
2180  REM * ARRAY NOW SORTED SO
2190  RETURN
3000  REM * SUBROUTINE TO CALCULATE THE VARIOUS
      STATISTICS
3010  REM * FIRST THE MEAN, VARIANCE, S.D. AND M.D.
3020  REM *   TX = RUNNING TOTAL OF VALUES
3030  REM *    N = NUMBER OF VALUES
3040  REM *   TD = SUM OF ABSOLUTE DEVIATIONS
3050  REM *  TD2 = SUM OF DEVIATIONS SQUARED
3060  REM *    M = MEAN
3070  REM *   VA = VARIANCE
3080  REM *   SD = STANDARD DEVIATION
3090  REM *   MD = MEAN DEVIATION
3100  LET TX = 0
3110  REM * INITIALIZE COUNTER
3120  FOR X = 1 TO N
3130  LET TX = TX + D(X)
3140  REM * RUNNING TOTAL
3150  NEXT X
3160  LET M = TX/N
3170  REM * THE MEAN
3200  LET TD = 0
3210  LET TD2 = 0
3220  REM * INITIALIZE COUNTERS
3230  FOR X = 1 TO N
3240  LET D = M - D(X)
3250  REM * D = DEVIATION FROM THE MEAN
3260  LET TD = TD + ABS(D)
3270  REM * ABSOLUTE (POSITIVE) DEVIATION
3280  LET TD2 = TD2 + D*D
```

```
3290  REM * SQUARE OF DEVIATION
3300  NEXTX
3310  LET MD = TD/N
3320  LET VA = TD2/N
3330  LET SD = SQR(VA)
3400  REM * NOW WORK OUT THE MEDIAN
3410  REM * QUARTILES AND VARIOUS RANGES
3420  REM * ME = MEDIAN
3430  REM * LQ = LOWER QUARTILE
3440  REM * UQ = UPPER QUARTILE
3450  REM * RA = RANGE
3460  REM * IR = INTERQUARTILE RANGE
3470  REM * SI = SEMI–I.Q.R.
3480  REM * FIRST THE MEDIAN
3500  IF N/2 <> INT (N/2) THEN LET ME = D((N+1)/2)
3510  REM * ODD NUMBER OF VALUES
3520  IF N/2 = INT(N/2) THEN LET ME = (D(N/2) + D(N/2 + 1))/2
3530  REM * EVEN NUMBER OF VALUES
3540  LET L = INT(N/4 + 0.5)
3550  LET U = INT(3*N/4 + 0.5)
3560  REM * POSITIONS OF QUARTILES
3570  LET LQ = D(L)
3580  LET UQ = D(U)
3590  REM * ASSIGN QUARTILES
3600  LET RA = D(N) – D(1)
3610  REM * DATA SORTED SO RANGE = LAST – FIRST
3620  LET IR = UQ – LQ
3630  LET SI = IR/2
3640  REM * THE OTHER RANGES
3700  REM * ALL DONE SO
3710  RETURN
4000  REM * SUBROUTINE TO OUTPUT THE RESULTS
4010  REM * FIRST THE SORTED DATA
4020  FOR X = 1 TO N
4030  PRINT X,D(X)
4040  NEXT X
4050  PRINT:PRINT"NUMBER OF ITEMS = ";N
4060  PRINT"MEAN =                    ";M
```

```
4070 PRINT"VARIANCE=                   ";VA
4080 PRINT"STANDARD DEVIATION =  ";SD
4090 PRINT"MEAN DEVIATION =       ";MD
4100 PRINT"MEDIAN=                  ";ME
4110 PRINT"LOWER QUARTILE =       ";LQ
4120 PRINT"UPPER QUARTILE =       ";UQ
4130 PRINT"RANGE=                   ";RA
4140 PRINT"INTERQUARTILE RANGE =  ";IR
4150 PRINT"SEMI–I.Q.R. =            ";SI
4160 RETURN
```

This rather long program will calculate all of the measures of location and dispersion that we have met, with the exception of the mode. I have written it in a structured way so that if you want to amend it you need only work on one section.

There is considerable variation in the command used to get data from the keyboard. In this program I have used INKEY$ (line 1460). For other machines this will need to be amended (1460 A$ = GET$ on the BBC 'B' for instance).

Solutions to Exercises

Exercise 2.1 (p. 22)

1. *Results of fishing competition.*

Weight of fish (nearest 100g)	Frequency
1.0	1
1.1	3
1.2	5
1.3	4
1.4	2
1.5	3
1.6	9
1.7	6
1.8	2
1.9	5
2.0	3
2.1	3
2.2	1
Total	47

2. *Energy yields for soya bean samples*

Calories per 100g (to one d.p.)	Frequency
150–	6
160–	14
170–	12
180–	19
190–	10
200–	3
Total	64

3. (a) 1.95 – usual rounding.

(b) 10 years 0 months – ages always round down.

(c) 355cm – usual rounding.

Exercise 2.3 (p. 34)

1.

Class size	Frequency density (per cent per pupil)
0–15	0.175
16–20	0.84
21–25	1.66
26–30	2.94
31–35	6.38
36–45	2.96
46–60	0.57

Note: for 0–15 class, class length = 16 pupils (from 0 up to 15 inclusive).

2.

Specific gravity (s.g.)	Frequency
990–992	9
993–994	14
995	22
996	28
997–999	18
1000–1002	9
Total	100

73 per cent had an s.g. less than 996.5.

3.

Depth (inches to one d.p.)	Frequency density (days per inch)
0–0.5	189.1
0.6–1.0	246
1.1–1.5	168
1.6–2.5	23
2.6–3.5	9
3.6–5.0	14.7

Note: for 0–0.5 class, class length = 0.55 – 0 = 0.55 in.

Exercise 2.6 (p. 47)

1.

Place of origin	Angle (°)
UK	138
W. Europe	102
E. Europe	12
Japan	90
Others	18

2. Angles: 166°, 54°, 14°, 22°, 79°, 25°.
3. Angles: Five years ago 121°, 82°, 139°, 18°.
   Present 129°, 103°, 93°, 36°.

Areas should be in ratio 1.4:1.

Exercise 3 (p. 54)

1. Rectangular.
2. Negatively skewed (or J-shaped).
3. Normal.
4. J-shaped.
5. Positively skewed.

Exercise 4.1 (p. 57)

1.

Number	Frequency
1	0
2	5
3	7
4	6
5	1
6	8
7	2
8	12
9	3
10	6
Total	50

Modal number = 8.

2.

Head sizes (nearest cm)	Frequency
50	1
51	4
52	1
53	3
54	8
55	5
56	6
57	5
58	4
59	3
Total	40

Mode = 54cm.

3. Mode of Group A = 3; mode of Group B = 3
One ought to consider the whole distribution and not just the mode.

Exercise 4.2 (p. 59)

1. 1.75.
2. 6.85.

Exercise 4.3 (p. 62)

1. 1 002 millibars.
2. 10st 0.6lb.

Exercise 4.4 (p. 64)

1. 24.675 years.
2. 6.1.
3. 55.05cm.
4. Mean A = 3.5; mean B = 4.8.

Exercise 4.5 (p. 66)

1. (a) 2.6; (b) 6.1; (c) 53.3.

Exercise 4.6 (p. 69)

1. 1 002 millibars.
2. 10st 0.5lb.
3. 6.
4. 55cm.
5. Median A = 3; median B = 4.5.
6. Gus would use the mode. There is not, in this case, much to choose between the other two.

Exercise 4.7 (p. 72)

1. (a) Modal class = 2.5 – 5.5mm.
    (b) Median = 7.7mm.
    (c) Mean = 9.2mm.

2.

Weight (nearest kg)	Frequency
44–46	2
47–49	7
50–52	10
53–55	12
56–58	6
59–61	13
62–64	5
65–67	4
68–70	1
Total	60

Estimated mean = 55.85 kg.
Estimated median = 55.45 kg.
Actual mean = 55.95 kg.
Actual median = 55 kg.

Note: since the original measurements were correct only to the nearest kg, the 'Actual' mean and median are, themselves, only estimates.

3. Mean = 26.7 years (I took the last class as 50–99)
Median = 24.6 years.

Exercise 5.1 (p. 77)

1. 0.32lb.
2. 4.3.
3. Mean = 15.07in., mean deviation = 0.22in.

Exercise 5.2 (p. 80)

1. 0.39lb.
2. 0.25in.
3. Mean = 10.6%,
mean deviation = 1.2%,
standard deviation = 1.3%.

Exercise 5.3 (p. 81)

1. Kendall Rd: 32.4 years, Gosset Ave: 15.1 years.
2. $m = 10.0$mm,
$s = 0.2$mm,
mean deviation from median = 0.1mm.
3. $m = $ N\$13 246, $s = $ N\$3 624
(I took N\$25 000 as the top limit).

Exercise 6.1 (p. 86)

1. (a) 5/26; (b) 4/11; (c) 0.
2. (a) 12/52=3/13; (b) 2/52=1/26; (c) 12/52=3/13.
3. (a) 3/6 = 1/2; (b) 2/6 = 1/3; (c) 5/6; (d) 1.
4. (a) 20/50 = 2/5; (b) 45/49.

Exercise 6.2 (p. 89)

1. 3/8.
2. (a) 1/16; (b) 4/16 = 1/4; (c) 12/16 = 3/4.
3. (a) 6/64 = 3/32; (b) 32/64 = 1/2; (c) 10/64 = 5/32; (d) 15/64.

Exercise 6.3 (p. 94)

1. (a) 27/64; (b) 1/64; (c) 37/64.
2. (a) 17/81 = 0.210; (b) 584/729 = 0.801.
3. 544/625 = 0.870.
4. (a) 6/720 = 1/120; (b) 180/720 = 1/4;
    (c) 210/720 = 7/24; (d) 48/720 = 1/15.

Exercise 6.4 (p. 96)

1. (a) (i) 15/25 = 3/5; (ii) 6/25; (iii) 4/25;
    (b) 9261/15625 = 0.593; (c) 27/125.
2. (a) 250; (b) 250/500 = 0.5; (c) 290/500 = 0.58;
    (d) 29/100 = 0.29; (e) 94/500 = 0.188; (f) Yes.

Exercise 6.5 (p. 99)

1. (a) 3/51 = 1/17; (b) 48/51 = 16/17; (c) 4/51; (d) 47/51;
    (e) 12/2652 = 0.0045; (f) 192/2652 = 0.0724;
    (g) 384/2652 = 0.1448.
2. (a) 120/1000 = 0.12; (b) 400/1000 = 0.4;
    (c) 70/1000 = 0.07.
    They seem to be dependent since 0.12 × 0.4 = 0.048 which is quite a lot less than 0.07.

Exercise 7.1 (p. 102)

1.   (i) (a) 125;(b) 80.
     (ii) (a) 78.1;(b) 128.
    (iii) (a) 120;(b) 83.3.
    (iv) (a) 91.3;(b) 109.6.
     (v) (a) 277.8;(b) 36.

Exercise 7.2 (p. 104)

1. (a)

	1903	1904	1905	1906	1907
(i)	76.9	84.6	100	93.3	103.8
(ii)	74.1	81.5	96.3	88.9	100
(b)	100	110	118.2	92.3	112.5

2.

	1905	1906	1907	1908	1909	1910
(a)	100	85	79.1	83.8	81.3	80.5
(b)	124.3	105.6	98.2	104.1	101.0	100

Exercise 7.3 (p. 106)

1. John, 44; Jane, 46; Claude, 74; Cecilia, 49.
2. A, 27; B, 38; C, 22.

Exercise 7.4 (p. 108)

1. 100,119.
2. (a) 224.3; (b) 300.
3. 110.8.
4. $x = 7$.

Exercise 7.5 (p. 113)

1. (a) 9.63, 16.13; (b) A: 11.292, 0.893, 2.480, 3.081, 3.007, 9.660, 22.267, 83.871. B: 13.441, 0.753, 3.386, 2.616, 2.623, 9.649, 22.797, 48.980; (c) 13.15, 10.35.
2. Sixty-eight live births per 1000. It does not deserve its reputation.

Exercise 8.1 (p. 123)

1. 0.929.
2. −0.9. Strong negative correlation implies that money does not necessarily buy success.
3. 0.9. There is a strong connection between performance in Maths and Physics.
4. 0.2; 0.6; −0.14. Not really.

Exercise 8.2 (p. 128)

1. (a) $y = 2x + 1$; (b) $y = 7x − 7$;
   (c) $y = 2x − 239$; (d) $y = −2x + 843$;
   (e) $y = −2x + 11$; (f) $y = −2\frac{1}{3}x + 28\frac{1}{3}$.

Exercise 11.1 (p. 150)

	(a)	(b)	(c)
1.	4 250	4 200	4 000
2.	53.8	54	50
3.	468 000	470 000	500 000
4.	690	690	700
5.	2.01	2.0	2
6.	408	410	400
7.	3.14	3.1	3
8.	93 200 000	93 000 000	90 000 000

Exercise 11.2 (p. 151)

	(a)	(b)	(c)
1.	9.55	9.6	10
2.	998	1 000	1 000
3.	9830	9800	10 000
4.	0.992	0.99	1
5.	64.8	65	60
6.	0.00103	0.0010	0.001
7.	0.00000465	0.0000047	0.000005
8.	0.0000985	0.000098	0.0001

Exercise 11.3 (p. 152)

	(a)	(b)	(c)
1.	3.142	3.14	3.1
2.	9.101	9.10	9.1
3.	0.699	0.70	0.7
4.	0.472	0.47	0.5
5.	2.172	2.17	2.2
6.	412.369	412.37	412.4

Exercise 11.4 (p. 154)

1. Max. = 61.8g, 15.45g; min. = 61.4g, 15.35g;
   absolute error = 0.05; relative error = 0.3 per cent.
2. 57.25; 25.8 per cent.